# Performance Appraisals within Public Safety

by

## Linda A. Sandleben

ISBN: 1-58112- 269-1

**DISSERTATION.COM**

Boca Raton, Florida
USA • 2005

*Performance Appraisals within Public Safety*

Dissertation.com
Boca Raton, Florida
USA • 2005

ISBN: 1-58112- 269-1

Dedicated to my Mother & Father, Jane & Joe, for all of the support and confidence they have had in me during my studies.

Table Of Contents

Performance Appraisals within Public Safety

Chapter I

## Introduction

Relative to other promotion methods, promotability ratings are seldom used in public safety organizations, despite their potential validity and low cost. To provide information for promotion decisions to fill higher positions in a state patrol agency, peer and higher-officer ratings and nominations were obtained for 60 lieutenants through majors. Results indicated this evaluation process was reliable and valid, providing sufficient variance for administrative decisions. Most participants were accepting of the process and results. Subsequent promotions have generally reflected the results. Sources of resistance to promotability ratings are discussed and suggestions made for dealing with this resistance. [1]

Public safety organizations have long used a variety of formal methods to determine promotion of officers, most often including written examinations, oral interviews, and assessment centers. One method that has seldom been used is promotability ratings, ratings of how well the officer will likely perform at the next level. A 2000 survey of 149 city, county, and state police organizations found that only 6 percent of these organizations used promotability ratings for promotion to sergeant through captain, whereas 90 percent used written exams,

44 percent used oral interviews, and 32 percent used assessment centers. [2]

In recent years, assessment centers have increasingly been used for promotions, and abundant research has conclusively demonstrated their validity. However, a recent meta-analysis of the comparative validities of promotional methods used in a variety of industrial and governmental organizations found that peer and supervisor ratings were slightly more valid predictors of future job performance than assessment centers. Also, peer and supervisor are more quickly and inexpensively obtained than developing and conducting an assessment center. Thus peer and supervisor ratings would seem to warrant more use for promotion in public safety organizations than they are currently receiving.

In the traditional performance appraisal process at the Office of Personnel Management (OPM), supervisors and subordinates develop a work plan at the beginning of the performance cycle, the subordinate carries out the work plan, the supervisor provides a midcycle review, and, at the end of the cycle, the supervisor provides feedback indicating how well the subordinate has fulfilled his or her work plan. However, with an increased emphasis on Total Quality Management (TQM), teamwork, and customer service, the supervisor may not have all of the information needed to provide complete and accurate feedback to the subordinate. [6]

If the individual is a manager who supervises others, these subordinates may be the best source of information on delegation, communication, and leadership skills. Peers may be in the best position to provide feedback on skills such as working with others, decision-making and technical capability. Finally, customers may be the best source of input on quality of work and service orientation. For these reasons, many organizations in the private and public sectors are incorporating information collected from multiple sources (supervisors, subordinates, peers, and customers) into the performance appraisal process.

This fact was recognized by a group of Career Entry Group (CEG) executives within OPM who asked the Office of Personnel Research and Development (PRD) to draft a proposal for incorporating subordinate, peer, and customer feedback into CEG's performance appraisal process. This paper presents the recommendations that were made for incorporating other sources of feedback into the performance appraisal process. The goal of the project was to improve the quality of information CEG employees receive about their job performance and, ultimately, to enhance the quality culture and level of employee involvement within CEG. [5]

The next section outlines, in more detail, the research process used to develop the recommendations, and is followed by a brief description of the current performance appraisal process used in CEG. Recommendations are presented in the final section of the paper. The paper also contains two appendices that provide more detail on the information collected from the research literature and from interviews with individuals in other public and private sector organizations.

A distinction is made here between peer/supervisor ratings of job performance, and peer/supervisor ratings of promotability. The latter are ratings of individuals' likelihood of performing well in higher-level jobs, based on their current performance on job dimensions which are important in higher-level jobs. The focus of promotability ratings is somewhat different that ratings strictly of current job performance, and arguably more valid to the extent that the higher-level jobs differ from the current level jobs. This would particularly be the case in moving from a basic officer level to the sergeant or first-line supervisor position, but also in moving from a supervisor position to a higher-level management position, typically involving more administrative and planning responsibilities. This article will describe the use of promotability ratings in a law enforcement agency, and will discuss several issues involved implementing this method. [3]

Personnel managers at public safety organizations play a key role in developing and implementing personnel policies and practices. While their influence varies from municipality to municipality and from issue to issue, the municipal personnel manager is an important source of expertise and knowledge concerning most human resource management issues.(1) A very important responsibility for many personnelists is performance appraisal system in public safety organization design and implementation. Performance appraisal is one of the most widely employed tools and is used for a variety of developmental and administrative purposes.(2) Performance appraisal is, however, a complex process and the center of considerable controversy over its utility and effectiveness. [21]

Proponents of Total Quality Management and other critics argue that individual performance appraisal engenders dysfunctional employee conflict, assumes a false degree of measurement accuracy, underemphasizes the importance of the work group in facilitating effective performance, assigns an inordinate amount of responsibility for poor performance to the individual employee, assumes the role of a hierarchical command and control device, and is subject to a whole host of heuristical and attributional errors.(3) A discussion of the merits of these critiques is beyond the scope of this article, but for those who value performance appraisal, this study is predicated on the belief that a properly designed appraisal system that is congruent with organizational culture can make an important contribution to organizational effectiveness.(4)

There are many rules, regulations, and guidelines associated with the performance appraisal process in the public safety organizations. Although most of them had no bearing on this project, a few key regulations could have potentially impacted how subordinate, peer, and customer feedback are incorporated into the performance appraisal system in CEG. For example, the final rating for each employee must come from one rating official. In addition, performance plans and ratings must be reviewed and approved by a higher level official for some employees. Finally, in some cases, performance elements and standards must be reviewed and approved by the union. [46]

It should be noted that the current performance appraisal process did not specifically call for subordinate, peer, or customer feedback to be taken into consideration when rating an employee. However, there was nothing in the regulations that prevented rating officials from collecting and considering this information when developing their final ratings. In fact, OPM guidance encourages managers to use additional information in the development of performance appraisal ratings.

Proponents of performance appraisal claim that an effective performance appraisal system in public safety organization produces specific performance feedback to improve employee performance, identifies employee training requirements, and links performance and personnel decision making.(5) The ultimate goal of most performance appraisal system in public safety organizations is to increase employee motivation and productivity. The recent research literature has begun to identify the characteristics of effective performance appraisal system in public safety organizations.(6)

Performance appraisal systems are an important tool of personnel management, but not all municipalities utilize this technique. The study analyzed the major reasons why public safetys do not implement a formal performance appraisal system. The data was derived from a national sample of public safety personnel officers. The results indicate that the most important barriers were a lack of expertise in performance appraisal and insufficient resources to adequately support a system. [50]

Discriminate analysis showed that western and southern cities were more likely to utilize an appraisal system, while eastern cities were least likely. Other significant variables associated with existence of a performance appraisal system were a favorable labor relations climate and lower turnover rates. Implications for development of a performance appraisal system are discussed.

Performance appraisal is an important tool of personnel management. The evaluation of employee performance is an ubiquitous process that occurs both formally and informally, irrespective of the existence of a formal performance appraisal system. Performance appraisal has had a controversial history with both strong proponents and detractors. Opponents argue that traditional individual performance appraisal assumes a false degree of measurement accuracy, ignores and devalues group performance, has a built-in conflict between employee evaluation and counseling, and engenders dysfunctional conflict and competition. Supporters argue that employees believe that personnel decision making should be based upon performance, that employees actively seek diagnostic and evaluative performance feedback, assert that performance can be measured accurately on some jobs, observe that jobs vary on whether performance is a product of an individual or a group and that greater employee participation can mitigate the conflict and defensiveness of traditional performance appraisal. It is beyond the scope of this article to analyze these arguments in detail, except to state that there is considerable merit on both sides. It is clear that an organization can operate with or without a formal performance appraisal system. A survey by Roberts (2000) found that approximately one in four public safetys do not operate a formal performance appraisal system while Daley (2001) found that one-sixth of North Carolina municipalities have no formal system. [28]

There has been little or no research devoted to studying the factors that influence the decision of a municipality to implement a performance appraisal system. There have been a number of recent surveys of public safety performance appraisal systems. None of these surveys have reported data on the number of municipalities that do not have an appraisal system and the reasons why. This article will explore some of the explanations offered as to why public safetys choose not to deploy a system and to identify the variables that statistically discriminate between municipalities that have systems versus those that do not. [9]


## Objective Of The Study

The primary objective of this study is to determine whether practicing personnelists and research scholars agree on what constitutes an effective performance appraisal system in public safety organization. If Personnel managers at public safety organization possess knowledge of the characteristics of effective performance appraisal system in public safety organizations, they are more likely to design and/or advocate performance appraisal system in public safety organizations that include these important attributes. In addition, the research is an opportunity for the academic community to learn from the experience of practitioners. Concepts and techniques that are valued by academics may not be functional or effective for those involved in the development and administration of performance appraisal system in public safety organizations. Thus, areas of disagreement can serve to identify subjects for further research and improve both theory and practice.

Importance Of The Study

Research of this genre is also necessary because much performance appraisal research is laboratory based with consequent questionable external validity. This research and other recent works attempt to delineate the contextual factors that affect the operation of performance appraisal system in public safety organizations.

Given the multitude of variables that can influence the development and administration of a performance appraisal system in public safety organization, personnel manager knowledge alone cannot be expected to have a significant impact on a given systems' effectiveness. Knowledge of the elements of an effective system is a necessary, albeit not a sufficient, condition to ensure performance appraisal system in public safety organization effectiveness.

How can Personnel managers at public safety organizations develop their knowledge of effective appraisal system practices? Formal education is one avenue. Personnel managers at public safety organization with advanced degrees specializing in personnel areas are more likely to be cognizant of the requirements for effective performance appraisal system in public safety organizations. Knowledge about performance appraisal can be obtained through specialized courses and training seminars, membership in professional associations, and self-study of the performance appraisal literature. Finally, practical knowledge gained from direct experience with performance appraisal is likely to be extremely important, if not the most important, influence. [12]

There are no published works that compare the opinions of personnelists with views of research scholars on the attributes of an effective performance appraisal system in public safety organization. Most of the published works have been descriptive or case study oriented and have not addressed the full-range of attributes related to appraisal system success.(7)

## Options for the Appraisal

1. Annual Job Performance Appraisals

In effect for many years, these appraisals guarantee communication between the superior and subordinate regarding the subordinate's performance. However, they are very limited for comparing many individuals due to the highly questionable reliability of ratings by only one rater (the immediate supervisor) for each officer.

2. Assessment Centers

On two occasions in the past ten years, lieutenants and above of the WSP have been appraised by an assessment center. However, the assessment centers were found to be expensive; reactions of the participants were less than enthusiastic; and participants did not feel that the results of the assessment center were used in making promotion decisions. For there reasons, there was little support for another assessment center.

3. Promotability Ratings

Both higher-ranking officers and peers are in an excellent position to evaluate the promotability of officers. In addition to the immediate supervisor, other higher-ranking officers are naturally observing officers' performance in a variety of forms - e.g., written reports, officers' judgment and thinking in planning sessions, and their interaction with other employees. Traditionally, higher officers would be expected to rate lower-ranking officers' promotability. One police promotability ratings process, using higher-ranking raters, has been described by Scott (2001), and a similar process for a fire department has been reported by Davidson (1999). [19]

The Problem

Two years ago, the Washington State Patrol (WSP) was faced with the likelihood that several higher-ranked individuals (captains and above) would retire in the next 12 to 24 months, requiring promotion of lower-ranked individuals to fill the higher positions. The Chief of the WSP has the authority to promote lieutenants and above as he sees fit, and traditionally, the Chief has made these promotion decisions based on his own knowledge of eligible individuals. However, it was felt that a systematic evaluation of the promotability of individual lieutenants and above could assist the Chief in making these promotion decisions, as well as decisions about the training/development of individual officers.

In this agency, promotion of troopers and sergeants is determined by a thorough, systematic appraisal of these officers' capabilities for handling higher positions. For sergeants, this appraisal consists of written examinations, an oral examination, a scored resume, command ratings, and forced choice evaluations of promotional potential, with each of these components weighted according to state statute. Thus, the WSP routinely used and has been accustomed to forma, systematic appraisals for promotion of lower-ranked individuals, but not for promotion of mid-level and higher-ranked individuals. A formal system was not to be developed for appraising the promotability of 35 lieutenants, 19 captains, and 6 majors.

## Hypothesis

Hypothesis 1: Breadth of knowledge will be positively related to safety performance.

Hypothesis 2: Depth of knowledge and skill will be positively related to safety performance.

Hypothesis 3: Depth of knowledge and skill will be more strongly related to safety performance than breadth of knowledge.

Chapter II

## Literature Review

Peers are also a potential source of promotability ratings. Peers are aware of the demands of higher-level positions and usually view each other in more familiar light, and in different work situations, than those observed by higher officers. Peer ratings have seldom been used in law enforcement agencies. A 2001 WSP survey of state police agencies' promotion methods found no state using peer ratings. However, reviews of research on the use and validity of peer ratings in mainly military and industry settings have been very positive. One of the few studies of peer ratings in a law enforcement agency found peer ratings (conducted for research purposes) correlated significantly with assessment center results for sergeants (r=.45) and lieutenants (r=.74). [25]

It was felt that obtaining and combining higher-officer and peer ratings would be useful for several reasons: providing a larger number of raters and, therefore, more reliable measures, combining different perspectives of performance; and allowing one source of ratings to offset possible bias in the other source. There was precedent within the WSP for using both supervisor and peer ratings: both have been routinely obtained to effectively validate the evaluation system used to promote troopers and sergeants, and command ratings are a small, but well-accepted component in the promotion of sergeants.

It was therefore decided to the higher-officer and peer ratings for this evaluation of mid-level and higher officers, due to the precedent of satisfactory prior use of higher-officer ratings for promotion of lower-rank officers in the organization, the demonstrated validity of peer ratings in other settings, anticipated high reliability due to combining higher-officer and peer ratings, and the practical advantages of obtaining these ratings quickly and at low cost. The effectiveness of the process was to be evaluated by the dispersion of ratings, the ratings' reliability and validity, participants' reactions, and practical use of the ratings. [47]

There are a number of reasons why a public safety would not deploy a performance appraisal system. A performance appraisal system requires a considerable investment in time, energy, and fiscal resources. For example, an effective system requires that specific performance standards, the rating form, administrative procedures and performance goals be developed. It is essential that raters are trained and employees oriented. Effective documentation of performance requires a considerable amount of the rater's time. The burdens on raters are exacerbated when supervisors rate large numbers of employees, lack the expertise to rate employee performance, or are unable to observe a representative sample of employee performance. A complete performance appraisal program requires a systematic evaluation by analyzing the reliability and validity of performance ratings to identify biased or substandard raters. It requires a careful assessment of user attitudes to identify specific problem areas. All of these steps demand a considerable investment, an investment that many governments may be unable or unwilling to make. [75]

A second factor that could impede the development of a performance appraisal system is a lack of expertise. The performance appraisal process is an extremely complex set of tasks that requires specialized knowledge in psychometrics, performance measurement, and highly developed interpersonal skills including the provision of positive and negative feedback. It requires cognizance of the cognitive process underlying performance appraisal including how raters process information and the preferred method (diary keeping) for maintaining comprehensive documentation (Ilgen & Feldman, 2001; Greenberg, 2000). Personnel managers may lack the expertise necessary to design an effective system and outside consultants are expensive, hence the lack of expertise may dissuade a municipality from developing a performance appraisal system. [104]

A third factor may be union opposition. Unions traditionally have emphasized seniority as the preferred basis for personnel decision making. Seniority is established on the assumption that experienced workers are more proficient and productive. It also emphasizes rewards for loyalty based upon years of service. Seniority is promoted by unions because they distrust management's ability or motivation to make unbiased performance ratings and personnel decisions.

Upper-level support and commitment is essential for the success of most managerial innovations. If the city manager, council or mayor opposes the concept or practice of performance assessment, it is unlikely that a performance appraisal system will be adopted. Upper-level support in performance appraisal requires an investment of resources, tangible and symbolic support of the concept, and evaluating raters on how well they administer the process. [175]

The final two explanations involve user attitudes toward performance appraisal, especially acceptance. In order for any appraisal system to be effective, both raters and ratees must accept the system. Employees may resist performance appraisal for a variety of reasons. [132] They include a lack of objective performance measures, the inability to assess individual performance because the work process is group oriented, perceived supervisor bias (i.e., by race, sex, affective orientation, politics), value conflict (i.e., performance rating should be for counseling purposes only) and a perceived lack of utility. Perceptions of a lack of utility occur frequently in situations where personnel decisions are made mostly by seniority. Thus, the expectancy that performance and rewards will be linked is low or zero, hence performance appraisal may not motivate behavior. [168]

Within the management and organizational behavior literatures, the linkage between organizational practices and individual performance is generally conceptualized and studied as follows: Societal and organizationally espoused values will engender particular types of general and human resource management practices. [125] These management practices will in turn influence employee attitudes, perceptions of the work environment, and knowledge and skill acquisition; ultimately, these latter individual difference variables will directly impact individual performance. In these literatures, individual performance is most often viewed as synonymous with the actions that employees engage in or display. Although decades of theory and research have been devoted to conceptualizing and measuring antecedents of individual performance, individual performance as a construct domain has received very little theoretical attention or research.[186] Without a clearer understanding of the substantive dimensionality of individual performance--arguably the most critical individual difference domain in organizational behavior and human resource management--the study of predictors and casual models of individual performance will remain enigmatic.

Although there have been numerous calls over the last 50 years to spend as much time and energy theorizing and studying individual performance criteria as has been spent on the predictor side, we have only recently begun to understand how broadly defined individual performance should be for the purpose of studying predictor-criterion relationships. [191] Evidence from recent decades of validity generalization/meta-analysis research and several large scale selection test validation studies suggest that it is primarily at relatively broad analytic levels (i.e., with respect to broad groupings of jobs or job families) that we should expect certain individual attributes (i.e., cognitive abilities and personality variables) to differentially relate to individual performance. That is, the available empirical evidence indicates that theorizing and studying the dimensionality of job performance with respect to specific jobs or posit ions (for the subsequent purpose of examining individual antecedents of job performance) may not always be necessary. A viable conclusion is that general or generic modeling of the dimensionality of performance within a performance domain may be appropriate for both scientific and practice concerns. [135]

Several studies have provided empirical support for a common factor structure of job performance within a performance domain. For instance, Hunt explored the dimensionality of job performance across a wide range of entry-level jobs in samples drawn from retail stores, home improvement centers, supermarkets, drug stores, restaurants, and theatre chains. He asserted that actions, such as maintaining personal hygiene and treating coworkers civilly, are basic, core behaviors that employees are required to display to some degree in almost all entry-level, hourly wage jobs, such as kitchen workers, cashiers, stock handlers, and data processing clerks. [111] Hunt's study provided strong factor-analytic evidence for the multi-dimensional nature of general job performance across these diverse entry level jobs.

In a similar vein, employees are required to display core, basic safety behaviors to some degree across jobs in certain industries such as manufacturing, mining, chemical processing, nuclear power plant operations, municipal public services (fire, police, and emergency medical services), and others.[107] In these industries, there exist segments or groupings of jobs that require high levels of procedural knowledge and skill in order to perform safely, thus protecting the public, the environment, the organization, and the workers themselves.

In the growing body of research on behavioral aspects of safety, a number of researchers have measured general safety behaviors in different industries. For the most part, this literature has focused on employee safety compliance, the extent to which employees adhere to safety procedures and carry out work in a safe manner. With only a few exceptions, researchers measure safety performance with respect to a single, overall scale. It is significant that an examination of the dimensionality of safety performance within an industry has not appeared in the literature.[78] Understanding the meaning (dimensionality) of safety performance as well as the knowledge- and skill-related antecedents of safety performance is critical for guiding management efforts toward the enhancement of safe work behavior and the reduction of negative individual, organizational, and societal consequences of unsafe work behavior. This understanding is particularly important within high risk industries, such as nuclear power, where safe work behavior has broad consequences. [157]

The goal of this research is to develop and empirically evaluate a model of general safety performance that is potentially applicable to safety performance in many work domains. The remainder of the introduction unfolds as follows. First, we discuss basic assumptions concerning the measurement of safety performance. Subsequently, we propose an a priori 4-

factor model of general safety performance.[136] Finally, we provide information on the context for this research (i.e., a test of the 4-factor model within the domain of hazardous waste work) and present the purposes of two studies designed to assess the construct validity of the proposed 4-factor model of general safety performance. [115]

## Conceptualizing General Safety Performance

General safety performance is defined as the actions or behaviors that individuals exhibit in almost all jobs to promote the health and safety of workers, clients, the public, and the environment. [138] Several basic assumptions concerning the nature and measurement of general safety performance are made. First, consistent with the safety performance literature, we assume that general safety behaviors can be scaled with respect to the frequency that employees engage in the behaviors. Second, consistent with the job performance literature, we assume that safety behaviors covary in meaningful ways, yielding an interpretable, multidimensional (correlated) factor structure. Third, we assume that general safety performance factors are distinguishable in terms of their determinants (e.g., procedural knowledge and skill) and covariation with other variables (e.g., absences, accidents, and illnesses). These assumptions, along with reviews of the relevant literatures and a content-oriented measurement development study (to be discussed in the Method

section), led to the specification of the a priori 4-factor

model of general safety performance.[21] Below, we describe the

respective factors for this model and discuss distinctions

between these factors and similar factors in the safety

performance and safety climate literatures.

Conceptual definitions for the four factors comprising the

4-factor model of general safety performance. Using Personal

Protective Equipment, Engaging in Work Practices to Reduce Risk,

Communicating Health and Safety Information, and Exercising

Employee Rights and Responsibilities. The factors comprising

this model are consistent with performance constructs specified

in the job performance literature, the literature on safety

performance, the literature on safety training, and the

literature on safety climate and safety culture.[187] For instance,

the factor Engaging in Work Practices to Reduce Risk

conceptually overlaps with measures in the safety performance

literature emphasizing worker safety compliance.

Although the safety performance literature often considers

the proper use of personal protective equipment as an element of

safety compliance, we viewed Using Personal Protective Equipment

as an independent performance factor.[119] Within a large number of

industries, individuals receive extensive personal protective

equipment training that is believed to be essential to the

development of procedural knowledge and skills with respect to this performance dimension. For instance, within the nuclear waste cleanup industry, hazardous waste workers need to be proficient in the use of various types of equipment to protect themselves against three types of radioactive wastes: high level wastes, transuranic wastes (i.e., liquid and solid wastes contaminated with elements above uranium in the chemistry periodic table such as plutonium, americium, and neptunium), and low level solid and liquid wastes.[181] Therefore, given the criticality of properly using personal protective equipment and the unique, requisite procedural knowledge and skills for using such equipment in a number of industries, we conceptualized Using Personal Protective Equipment as an independent, yet correlated, dimension of general safety performance. [138]

The dimensions Communicating Health and Safety Information and Exercising Employee Rights and Responsibilities are conceptually similar to dimensions in the safety climate literature focusing on safety communication, Mearns, O'Connor, & Bryden, 2000) and effective reporting of incidents/accidents, respectively. Although these latter climate dimensions are somewhat similar to two of the hypothesized safety performance dimensions, these safety climate dimensions and other climate dimensions are often operationalized with respect to employee perceptions of organizational safety policies and management safety practices or in terms of employee satisfaction with the respective policies and practices.[165] That is, in contrast to the conceptualization and measurement of safety performance factors as actions or behaviors that employees exhibit, the focus of safety climate is not on particular behaviors of employees. We should also note that the factor Communicating Health and Safety Information is consistent with conceptual factors of general job performance concerning communication. [189]

Context and Purposes of the Overall Study

The purpose of Study 1 was to empirically evaluate the a priori 4-factor model of general safety performance as a demonstration project within the hazardous waste work domain. That is, as an operationalization of the 4-factor model, the items serving as indicators of the four factors include a mix of mostly generic or general items that are applicable to many types of work involving safety issues and several items that are more context-specific (i.e., items that apply primarily to hazardous waste work). In Study 1, we present results of confirmatory factor analytic tests of the 4-factor model of general safety performance with coworker (peer) appraisals from employees in four organizations. The employees worked in 23 different jobs at the 560-square mile Hanford nuclear waste site situated along the Columbia River in Washington State. The

Hanford site is the most contaminated of the 14 large sites in 13 states that makeup the U.S. Nuclear Weapons Complex. [112]

Using the general safety performance factors confirmed in Study 1, in Study 2 we examined the extent to which increases in employees' breadth of knowledge (operationalized as the diversity of training with respect to a performance factor) and depth of knowledge (operationalized in terms of the amount of advanced or refresher training with respect to a performance factor) are predictive of supervisory ratings of safety performance. [178]

Study One

Method

Participants

The participants for Study 1 were 574 hazardous waste workers who provided anonymous ratings of the safety performance of their coworkers. The targets of the coworker appraisals were employees of three DOE contractors and one subcontractor. The targets of the performance appraisals needed to have minimally completed the general 40-hour HAZWOPER (Hazardous Waste Operation and Emergency Response) course and to be involved in some form of hazardous waste work at the Hanford nuclear site in Washington State. The jobs held by employees were representative of jobs at the Hanford site including nuclear waste process operator, health physics technician, electrician, field engineer, material handler, craft supervisor, and plumber/pipe fitter.

Four-factor general safety performance model and measure. The following is a summary of the background work on the development and administration of the general safety performance measure. First, an extensive review of over 200 articles in the job performance, safety performance, safety climate, and safety training literatures was conducted. In addition, we reviewed the lesson plans for general hazardous waste worker training as conducted by almost all major labor unions that represent hazardous waste workers (i.e., International Brotherhood of Teamsters, 2003; International Union of Operating Engineers, undated; National Ironworkers and Employees Apprenticeship Training and Journeyman Upgrading Fund, undated; Oil, Chemical, and Atomic Workers International Union--The Labor Institute, undated; and The United Brotherhood of Carpenters Health and Safety Fund of North America, 2002). This review indicated that although existing general job performance models, safety performance studies, an d other organizing frameworks for worker behaviors that included safety-related factors could assist in identifying preliminary behavioral performance constructs and items, these literatures were inadequate for comprehensively specifying the behavioral domain of safety performance and, in particular, hazardous waste worker performance. Therefore, a review of these literatures along with a content-oriented validation study led to the development of a preliminary 50-item safety performance measure.

Next, coworkers rated the frequency (using a 7-point scale ranging from never to always) with which hazardous waste workers exhibited each of the 50 behaviors. Third, item analyses and feedback meetings with line workers and groups of supervisors were conducted. Based on the results of the item analyses and feedback meetings, responses to the 27 items were retained for the present factor analyses and grouped into four a priori factors as described above. We should note that the eliminated behavioral statements were removed because it was found that they did not apply to a vast majority of employees or required respondents to rate the extent to which an employee had "demonstrated knowledge" of a particular subject matter.

Coworkers were asked to rate the safety performance of a "typical coworker." A typical coworker was defined as "the person that you usually work with or the average performer in your work group." This anonymous coworker rating procedure was necessary for ensuring the participation of various labor union members represented by the Hanford Atomic and Metals Trade Council. Coworker appraisals were obtained in group sessions, where the appraisers were informed that their evaluations would only be used for research purposes.

Specification and test of general safety performance model. The 4-factor model was specified by restricting each of the 27 items to load only on the factor it was hypothesized to represent. The analytic strategy consisted of three steps. First, cases in the database with a large percentage of missing data (i.e., 25% or greater) were eliminated. This brought the coworker sample size from 574 to 550. Second, in order to retain the maximum amount of data for the confirmatory factor analysis, missing data (typically data for one or two items for 249 cases) were estimated using a regression-based multiple imputation procedure developed by Schafer (1997). (1) Third, the confirmatory factor analytic test of the 4-factor model was conducted using LISREL 8. The fit of the model was judged with respect to the chi-square test, overall goodness of fit statistics (i.e., CFI and GFI), an analysis of residuals (i.e., RMSEA and Standardized RMSR), and the magnitudes of item factor loadings.

## Results and Discussion

The item means, item standard deviations, and factor-loading matrix for the 4-factor model. All factor loadings for the model were greater than or equal to .62 and statistically significant. The overall fit statistics for the 4-factor model (i.e., [chi square] [318, N = 550] = 1,377, p <.01; CFI .90; GFI = .84) and residual statistics (RMSEA .078; Standardized RMSR = .048) indicated that the model provided an acceptable fit to the data. The correlations among the four factors ranged from .62 to .79. In spite of high correlations among the four factors, the 4-factor model fit the data considerably better than a single first factor, representing a possible common methods factor ([chi square] [324, N = 550] = 2,295, p <.01; CFI = .81, GFI .71; RMSEA = .11, Standardized RMSR = .062).

Overall, the results from Study 1 provided support for a correlated, 4-factor model of general safety performance. Notably, the results indicated that the four factors were highly correlated, leaving open the possibility that a common higher order factor may underlie general safety performance. In addition, although the LISREL analyses identified several complex items that could be loaded on other factors or eliminated to improve the overall fit statistics for the 4-factor model, our decision was to retain all items as originally hypothesized. This decision yields a parsimonious general safety performance model that includes a comprehensive set of general behaviors representing this construct domain. In a subsequent section, we will discuss how the measure might be modified to measure general safety performance in other occupations and types of work.

As noted above, the four confirmed safety factors would be hypothesized to have different knowledge and skill determinants. Empirical support for hypothesized relationships between scores on safety dimensions and indicators of knowledge and skill constructs would provide further support for the construct validity of the 4-factor safety performance model. We now turn to Study 2, which was designed to test hypothesized relationships between knowledge and skill constructs and general safety performance factors.

## Study Two

In the fields of applied psychology and organizational behavior, performance is frequently expressed as a function of its determinants. A longstanding proposition has been that job performance is a multiplicative function of ability and motivation. This functional relationship, first specified by Maier (1999), has received a fair amount of research attention. The literature has produced mixed results concerning the interaction of ability and motivation in predicting performance. More recently, Sackett, Gruys, and Ellingson (1998) examined this multiplicative proposition when motivation was conceptualized and measured in terms of personality. Based on reanalysis of four large data sets, they concluded that ability-personality interactions are not detected at or above chance levels.

Campbell (2000) and Hunter and colleagues have proposed alternative models of performance determinants. Both sets of authors argue that declarative knowledge and procedural knowledge and skills (as indicated by work sample performance in Hunter and colleagues' research are direct antecedents of job performance. Declarative knowledge is defined as an understanding of the task requirements or the ability to state the facts, rules, and principles that are prerequisite for successful task performance. Procedural knowledge and skill is the capability attained when declarative knowledge (knowing what to do) is combined with knowing how and being able to perform a task.

Although the research of Hunter and his colleagues has focused on overall job performance as the criterion, Campbell's (2000) work has emphasized a taxonomy of performance factors as relevant criteria. An important feature of Campbell's (2000) framework is that knowledge (and measures thereof) should be isomorphic or similar in content with the respective performance factors.

However, to date, almost all research examining relationships between knowledge and job performance has linked overall job knowledge tests or composite measures of job knowledge with overall performance ratings or composite measures of job performance.

Importantly, declarative knowledge and procedural knowledge and skill are hypothesized within Campbell's (2000) model of performance determinants to be, in part, a function of education and training. In the hazardous waste worker domain, safety training is conducted in a manner that would be expected to enhance knowledge and skill acquisition and, thus, performance with respect to each factor. As a mandatory job requirement, all hazardous waste workers are initially required to take extensive general safety training. This training is highly structured and designed to not only provide trainees with a broad health and safety knowledge base with respect to central health and safety concepts, safety rules and regulations, and safe work behaviors, but also develop trainees' understanding of interrelationships among these factors. The training often incorporates multiple training methods including lecture and discussion, followed by behavioral modeling, and then practice in simulated contexts.

Increases in the diversity or breadth of the above types of training within a knowledge area (e.g., training related to the use of various types of personal protective equipment such as different types of respirators) would be primarily expected to enhance declarative knowledge (which we will refer to as breadth of knowledge). In particular, the enhancement of declarative knowledge would provide hazardous waste workers with a broad knowledge base for responding to new circumstances, changing situations, and emergencies. Novel and changing situations in hazardous waste work are similar to the notion of inconsistent tasks in the learning theory literature. Responding to inconsistent tasks is expected to be heavily dependent on one's available resources. Therefore, consistent with resource allocation models of skill acquisition and performance, we hypothesize that for each performance factor:

Hypothesis 1: Breadth of knowledge will be positively related to safety performance.

Increases in training in the form of safety refresher or recertification courses would be primarily expected to enhance procedural knowledge and skills through knowledge and skill updating and repeated-opportunities to practice with feedback. In this sense, skills are expected to become proceduralized or automatized with increased refresher instruction, leading to more consistent exhibition of routine safety behaviors on the job. We will refer to the mastery and maintenance of procedural knowledge and skills as depth of knowledge and skills. In terms of stage theories of learning, refresher training would be expected to enhance knowledge compilation, where individuals integrate sequences of cognitive and motor processes required to perform a task, and mastery of procedural skills. Furthermore, refresher training that exposes individuals to a variety of related or similar material across different presentations should permit individuals to more easily integrate the new materials with that already in memory. Yet, the mastery and maintenance of procedural knowledge and skill is expected to enhance the transfer of training to the job only when individuals have the opportunity to exhibit the knowledge and skills they have learned. In sum, consistent with stage theories of skill acquisition and performance and research on transfer of training, we hypothesize the following for each for each general safety performance factor:

Hypothesis 2: Depth of knowledge and skill will be positively related to safety performance.

In addition, given that general hazardous waste worker training involves knowledge and skill acquisition with respect to all safety performance factors, we also tested Hypothesis 2 with respect to depth of general health and safety knowledge and skill and each performance factor (and a composite measure of general safety performance). Furthermore, noting that depth of knowledge and skill with respect to general safety issues and the performance factors (i.e., Using Personal Protective Equipment, Engaging in Work Practices to Reduce Risk, Communicating Health and Safety Information, and Exercising

Employee Rights and Responsibilities) become translated into greater work experience over time, we will also test Hypothesis 2 with respect to a composite measure of depth of safety knowledge and skill and a composite measure of general safety performance. A test of the latter expectation will provide a linkage to the literature examining relationships between overall job knowledge tests or composite measures of job knowledge and overall performance ratings or composite measures of job performance.

Given that most general safety behaviors or tasks are consistent or routine tasks, we expect that depth of knowledge and skill will be more strongly related to performance on each factor than breadth of knowledge. In addition, over time, the number of safety refresher courses that one takes becomes translated into greater depth of work experience in a given safety performance area. Therefore, the expectation that depth of knowledge and skill will be more strongly related than breadth of knowledge to performance on each factor also is supported from work experience perspectives. Thus, for each safety performance factor, the third hypothesis is as follows:

Hypothesis 3: Depth of knowledge and skill will be more strongly related to safety performance than breadth of knowledge.

In addition, we will test Hypothesis 3 with composite measures of depth and breadth of knowledge and a composite measure of general safety performance.

Rater resistance is derived from a variety of sources. Raters may resist performance appraisal because the costs may exceed the benefits. Perceived costs include excessive time and energy devoted to system development and administration, disruption of amicable relations with employees, the inability to measure performance, a lack of efficacy in presenting performance feedback, and the inability to link performance and personnel decision making, among others. Raters sometimes use performance appraisal as a control device, and a system that can't be used for administrative decision making has little control value.

Chapter III

Method

Participants and Procedure

The participants were 35 lieutenants, 19 captains, 6 majors, and 4 deputy chiefs. All were advised in writing that each of the lieutenants, captains, and majors would be confidentially rated by peers and by higher-rank officers to assist the Chief in evaluating individuals for future promotion and training. Participants were requested to gather for the rating process at a designated time and place to ensure uniform instruction and orientation. In the orientation, participants were advised of the following:

1. They were to rate each other's promotability on each of four job dimensions: judgment/decisions, administrative skills, personal impact (leadership, "people skills," and communications), and work involvement (initiative, loyalty, adaptability, and handling stress). These dimensions were developed by a committee of two majors, one lieutenant, and the author by considering critical areas of the job of captain and above, and by a review of the dimensions used to appraise job performance of lieutenants and above. These dimensions were considered specific enough to focus raters on the main relevant skills and qualities needed by higher managers, but not too numerous to be tedious and cumbersome for the rating process.

2. For each individual, ratings would be averaged for each dimension and then combined into an overall average. These averages would be calculated separately from peers and from higher-rank officers, as well as combined. Each officer would receive his individual results, along with the relative position (top, middle, or bottom third) of his overall peer and higher-rank averages. All results would be given to the Chief and one deputy chief, but to nobody else. Each rater would sign the rating form, but these forms would be kept confidential.

3. Officers were instructed to rate only those individuals with whom they had frequent, recent job contacts. They were advised to rate an individual on only those dimensions in which they directly observed the individual perform, rather than on the bases of the individual's reputation.

4. Ratings would be on a scale of 1-5, with a "5" representing "an outstanding prospect, far exceeding normal requirements," a "3" representing "an average prospect, meeting normal requirements," and a "1" representing "an individual who is probably not promotable at this time."

Following this ratings, process, participants completed a second evaluation process, to supplement and compare against the results of the ratings. In the second evaluation, participants were asked to assume that the positions of the 6 majors and 4 deputy chiefs had just become vacant, and to nominate the lieutenants and above, including incumbents, they thought would do the best jobs in these generic positions. Due to the sensitivity of this exercise, participants were advised not to sign their nominations, and results would be provided only to the Chief and one deputy chief. Results would consist of the number of nominations separately, for major and deputy chief, and the combined nominations for each individual.

Data were collected from multiple sources, allowing for a number of different perspectives to be taken into consideration in the development of the recommendations. The data sources examined included:

* a review of the appropriate research literature. This included examining studies where subordinate, peer, or customer ratings had been used either in a research or organizational setting. Both technical articles and "how-to" articles were included in the review;

* telephone interviews with members of private and public sector organizations that have used subordinate, peer, or customer feedback in the performance appraisal process. Personnel Research and Development (PRD) psychologists identified a list of organizations conducting programs of this sort and contacted them first. Additional contacts were identified in these initial interviews. The telephone interviews were conducted by a Personnel Research Psychologist and two

managers in PRD. The interviews focused on the nature of the process in the organizations; who is providing feedback data; what areas of performance are being assessed; how data are collected, analyzed, and reported to employees; and how the feedback data are used;

    * an examination of the regulations and practices related to performance appraisal in the federal government;

    * focus groups and interviews with CEG management and employees. During the focus groups, participants were asked about the strengths and weaknesses of the current performance appraisal process, reactions to incorporating subordinate, peer, and customer feedback into the appraisal process, and, if implemented, how such a process could work best.

    In sum, the recommendations presented in this paper are based on information collected from a variety of sources, including past research, effective practices in private and public sector organizations, rules and regulations that cover performance appraisal in the federal government, and ideas and concerns of CEG employees and managers.

    The method employed to address the research objectives is a national mail survey of public safety Personnel managers at public safety organization. The International City Management Association (ICMA) provided a comprehensive listing of all public safetys stratified by population. Cities with populations over 10,000 were targeted because municipalities of this size were more likely to possess the resources and expertise to support a formal appraisal system. Questionnaires were mailed to 800 public safetys with populations greater than 10,000. The majority (750) were mailed to cities over 25,000. Cities were selected if there was a named personnel manager because research has shown that return rates are higher when the respondent can be identified.(8) Responses were received from 314 municipalities for a response rate of 39.25 percent. This response rate is congruent with other surveys of this type.(9)

The data was generated from a larger study of public safety performance appraisal system practices. The public safety performance appraisal system practices survey attempted to test specific hypotheses based upon a national sample of public safety performance appraisal system practices. Respondents were personnel professionals (mostly personnel managers) drawn from cities with populations greater than 10,000. Cities of this size were more likely to possess the resources, expertise and labor force to justify a performance appraisal system. The sample was generated from lists of personnel managers purchased from the International City Management Association. There were 800 questionnaires sent out, and 314 were returned for a response rate of 39.25 percent.

Measures of Depth of Knowledge and Skill, Breadth of Knowledge,
and Safety Performance

Depth of knowledge and skill: For each performance factor
(with the exception of Exercising Employee Rights and
Responsibilities), indicators of depth of knowledge and skill
were computed as the sum of the number of refresher training
courses taken and passed in the respective knowledge and skill
area. For instance, depth of knowledge and skill with respect to
Using Personal Protective Equipment was computed as the sum of
the number of times an individual had taken five equipment-
related refresher courses, such as the annual self-contained
breathing apparatus course and the quantitative mask fit
refresher course. In addition, indicators of depth of knowledge
and skill were computed for general health and safety training
as the sum of the number of general employee training courses
and 8-hour annual HAZWOPER training courses a worker had taken
and passed. Finally, an unweighted composite (sum) of the number
of refresher courses a worker had taken across knowledge and
skill areas was computed. We should note that the safety
refresher courses involved lecture presentations of new material
(e.g., legal developments, advances in equipment design)
followed in many cases by demonstrations and practice with
feedback.

Breadth of knowledge. For the performance factors, Using Personal Protective Equipment and Engaging in Work Practices to Reduce Risk, indicators of breadth of knowledge were computed by summing the number of different courses a worker had taken and passed in each area. For instance, breadth of knowledge concerning Engaging in Work Practices to Reduce Risk involved summing the number of different initial (nonrefresher) training courses an individual had taken dealing with confined space (2 possible courses), radiological control (12 possible courses), asbestos control (5 possible courses), lead control (3 possible courses), lock and tag (7 possible courses), packaging and transport (9 possible courses), and blood borne pathogens (1 possible course). These training courses often included a mix of lecture-based instruction and behavioral role modeling. An unweighted composite (sum) of the number of different initial (nonrefresher) training courses a worker had taken across knowledge and skill areas was also computed.

Breadth of knowledge could not be computed for knowledge related to Exercising Employee Rights and Responsibilities and Communicating Health and Safety Information. Material related to exercising one's rights and responsibilities is only presented in the general HAZWQPER course and there is only one specific hazard communication course.

Safety performance measures: The performance measures for Study 2 were based on the four confirmed factors of Study 1. Foremen who directly supervised the line employees provided the performance ratings in group sessions, where the appraisers were informed that their evaluations would only be used for research purposes.

We should note that we considered conducting a common factor analysis on the supervisory ratings to examine the invariance of the confirmed 4-factor model of safety performance. However, due to the pattern of missing data in the supervisory ratings of employee performance, we would have needed to impute data (on one or more items) for 79 of the 133 hazardous waste workers who were rated by their supervisor. Therefore, we relied on the confirmatory factor analyses of the coworker data for defining the performance scales (i.e., measures). However, statistical tests of differences between the coworker and supervisory means and standard deviations for the respective performance factors were not significant and both sets of distributions were negatively skewed. Performance on each general safety performance factor was computed as the average of the respective item ratings for the performance factor.

The findings from the overall study offer progress toward a taxonomy of general safety performance by directly addressing the criterion dimensionality issue. Four general safety performance factors were confirmed within Study 1 and labeled Using Personal Protective Equipment, Engaging in Work Practices to Reduce Risk, Communicating Health and Safety Information, and Exercising Employee Rights and Responsibilities. In addition to being grounded in the extant literature, the four safety performance factors are consistent with safety training content areas and lesson plans (in terms of being segmented into training modules) for a number of major labor unions. In essence, the four safety performance factors and behaviors that compose the domain of general safety performance are closely associated with an implicit theory of the safety performance domain. Importantly, the four general safety factors provide a starting point for future research on the dimensionality of general safety performance in other types of work.

Notably, the four general safety factors were highly correlated, with correlations among the factors ranging from .62 to .79 in Study 1 and from .51 to .74 in Study 2. This positive manifold, coupled with the fact that a test of a single, unitary factor model provided a poor fit to the data in Study 1, implies that a common higher order factor underlies the four first-order general safety performance factors. Along with a general lack of discriminate validity in the prediction of the four factors in Study 2, this positive manifold would suggest the need for

future research to address both the possible nature and presence
of a higher order safety performance factor. In effect, a
composite or overall safety performance score for the General
Safety --Performance Scale (GSS) may be meaningful as an
indicator of a higher order safety performance factor.

The degree to which the confirmed factors of the GSS
generalize to work domains other than nuclear waste is both a
theoretical and an empirical question. Due to similar actions
that workers engage in across many types of safety -related work
and the regulatory nature of safety -related work, we anticipate
that the four general safety factors would generalize to safety
behavior in other industries, such as agriculture, mining,
energy, manufacturing, construction, transportation, and public
services including health care. However, we expect that the four
subscales used in this study to measure the dimensions of
general safety performance will, in some cases, require slight
modification for use in other work domains. For optimal use in
other types of work, we suggest that researchers and
practitioners address the following three questions for the
respective items in each of the general safety performance
subscales: (a) does the item apply to this type of work (if not,
delete), (b) does the language of the item nee d to be modified
(if yes, modify accordingly), and (c) do items need to be added
to reflect relevant safe work behaviors in the work domain (if

yes, add appropriate items to the respective scales). Items that are more context-specific or items that may require editing prior to use in another type of work are identified.

As examples, several dimension-specific suggestions for modifying items of the GSS for use in other types of work axe as follows. First, Items 6 and 9 of Using Personal Protective Equipment may not apply to some types of work that are more individualized or work that does not require the use of pressurized air respirators, respectively. Second, Item 22 ("Contacts appropriate personnel when faced with questions and/or issues regarding HAZWOPER") would require minor editing. That is, the context-specific term HAZWOPER could be replaced with the word "safety." Third, an item such as, "Exercises rights to review MSDS and to use other reference materials that may provide additional health and safety information," could be reworded to include the particular safety manuals and reference guides for another type of work. Here, an MSDS (Materials Safety Data Sheet) may not be applicable to other types of work. Although the above recommendations may enhance the application of the GSS to other types of work, the above suggestions are not intended to preclude efforts to study more specific safety performance factors in either hazardous waste work or other types of work. Certainly, the possibility that more specific safety performance factors could assist in explaining safety behaviors in the present set of jobs or other jobs is very likely. For instance, recognizing and evaluating hazards and responding to emergencies are two possible performance factors that could be the subject of confirmatory factor analytic efforts to explain performance variability in other types of safety -related work, such as firefighting or emergency medical services. In addition, the study of contextually oriented performance in the safety domain would be a useful supplement to the present effort, which examined worker actions relative to more consistent, routine tasks.

Although a detailed discussion and empirical examination of the relation between general safety performance and routine task performance is beyond the scope of this study, the potential overlap between general safety performance and task performance is an important conceptual and practical issue. Across jobs, general safety performance and task performance are likely to overlap along a continuum. Within critical skills occupations where safety concerns are an important aspect of the work, the four dimensions that comprise general safety performance described here are likely reflected in task performance. Task performance has been defined as behavior involved in transforming raw materials into products and services or servicing or maintaining the technical core. General safety performance is inherent in these latter types of behaviors. For instance, task performance, such as operating a production machine in a manufacturing plant, will likely require engaging in certain actions, such as wearing any necessary personal protective equipment (Using Personal Protective Equipment dimension), properly disposing of any dangerous scrap material (Engaging in Work Practices to Reduce Risk dimension), and reporting any safety incidents/accidents while working on the machine (Communicating Health and Safety Information dimension). In many ways, the notion that safety performance is an inherent aspect of task performance in more safety -related types of work parallels Borman and Motowidlo's (2003) observation that contextual performance might be considered task (routine) performance in more service-oriented work. Indeed, safety behaviors have been included when measuring task performance. Another likely factor influencing the relationship between general safety and task performance is how effectiveness is defined, monitored, and rewarded with respect to task performance.

For instance, as is well documented in the safety literature, following safety procedures (e.g., taking the time to put on safety goggles) can add steps to the work process that can reduce efficiencies (e.g., how quickly a machine operator completes a work operation). Whether task performance is defined with safety considerations in mind is likely due to factors in the broader social environment like safety climate (e.g., Zohar, 2000) and supervisory messages relating to the important goals of the work group.

In addition to future research efforts aimed at clarifying the conceptual and empirical distinction between general safety performance factors and task performance factors, other research directions might include confirmatory factor analytic tests of safety performance factor structures with data from other sources (e.g., supervisors). Recall, the first study relied on employee ratings of the "typical" coworkers performance, which may have affected the attained factor solutions. In addition, examinations of other hypothesized individual difference antecedents of safety performance (ability and personality variables) including an assessment of whether or not these individual difference variables differentially predict the four dimensions of safety performance would further add to assessments of the construct validity of the GSS. Finally, an assessment of the degree to which individual difference - safety performance relationships are moderated by organizational climate would be a meaningful addition to the literature.
In regard to studying antecedents of safety performance, the present findings add to our understanding of how workers' training experiences contribute to safety performance. In particular, the present findings contribute to our limited

understanding of how breadth and depth of knowledge and skill relate to safety performance. The magnitudes and nonlinear nature of most relationships between depth of knowledge and skill and safety performance are consistent with relationships reported in the literature between job tenure, job knowledge, work sample performance, and overall supervisory performance ratings. In addition, the present linear depth of knowledge-safety performance relationships are likely lower than comparable estimates of job knowledge-performance relationships reported in the literature due to the fact that depth of knowledge and skill scores are restricted in range. If an individual completed a safety refresher course (i.e., updated his or her skills), then he or she would have exceeded a minimum passing score on a refresher course test. In addition, we believe that the moderate to high relationships between the various knowledge and skill measures are more reflective of the development of well organized safety knowledge structures and the fact that, over time, increases in safety knowledge and skill become closely associated with greater depth of work experience.

We should also note that the present observed safety knowledge - safety performance relationships are likely lower than might be found in other types of work as a result of substantive range restriction (and consequent negative skew) on the performance measures. That is, the distributions for the safety performance measures were restricted and negatively skewed, likely owing to the highly regulated nature of performance in the domain of hazardous waste work. Although our discussion to this point has focused on issues concerning the construct validity of general safety performance factors and breadth and depth of knowledge and skill variables, the present findings have several important practical implications. First, the magnitudes of the depth of knowledge-safety performance relationships indicate that the continuance of annual refresher training courses and advanced training is meaningfully related to safety performance. In terms of capitalizing on economies of scale, our findings support across-job safety training in the areas of general health and safety, personal protective equipment, and work practices aimed at reducing risk.

Although the four safety performance factors are highly correlated, another practical implication concerns the value of the 4-dimensional model in settings where specific aspects of safety performance are emphasized. Examples might include emphasizing the Using Personal Protective Equipment dimension to evaluate trainee safety performance following a training program for firefighters on the proper practices for using and inspecting their personal protective equipment. Alternatively, one could employ the Communicating Health and Safety Information dimension and measure to study the influence of leadership and organization support on safety communication behaviors. However, in other situations, a single composite measure may be more appropriate, for instance, when seeking to determine an employee's overall level of safety performance as one factor in conducting a performance evaluation, validating a selection procedure with respect to safety performance, or making a promotional decision.

Public safety performance appraisal systems were excluded because they traditionally are granted greater autonomy in personnel practices and personnel managers have less direct contact with police and fire performance appraisal systems. Of the 312 usable questionnaires returned, 72 (22.9 percent) did not administer a non- public safety performance appraisal system. It is probable that the population of cities without performance appraisal systems was higher because municipalities without performance appraisal systems had a lower incentive to respond.

Chapter IV

Results

Rating Process

Individuals' overall averages, across all dimensions and raters, ranged from a high of 14.4 to a low of 2.40, M = 3.27, SD = 0.41. Means and standard deviations of individuals' average ratings for each dimension, across rater, were as follow: Judgment/Decisions M = 3.20, SD = 0.47; Administrative Skills M = 3.37, SD = 0.48; Personal Impact M=3.34, SD = 0.46.

Each lieutenant/captain was rated by an average of 18.1 peers and by 11.1 higher officers, for a total average of 29.2 raters. Each major was rated by 5 peers and by 4 higher officers - 9 total raters. A high degree of agreement was found among rates:

1. Intraclass correlations for each dimension, which express the inter-rater reliability of ratings in each dimension, ranged fro, .90 to .93.

2. Overall average ratings made by peers and by higher-ranking raters correlated highly (r=.78, p <.001.)

3. A one-way ANOVA showed no significant mean differences in ratings made by peers vs. those made by higher-ranking raters.

A one-way ANOVA showed significant differences (F=15.79, p <.001) in the mean overall averages of ratees: majors-3.10, captains-3.39, and lieutenants-3.22. Considering that captains, as a group, are probably more qualified for higher positions than lieutenants, there figures seem to show a lack of bias based on higher rank and seniority. Interestingly, four of the seven more junior lieutenants placed in the top third of officers, based on overall ratings.

Also, there was no evidence of officers of one rank collectively rating themselves higher than other ranks. Majors rated captains and lieutenants higher (M=3.43 and 3.30, respectively) than they rated themselves (M = 3.18); captains rated themselves and lieutenants virtually the same (M = 3.26 and 3.25, respectively); and lieutenants rated themselves no higher (M = 3.25) than they were rated by the higher-rank officers.

Nominations Process

In this evaluation exercise, each lieutenant through major could potentially be nominated for major or for deputy chief by as many as all 64 lieutenants through deputy chiefs. The combined nominations for major and deputy chief for each individual ranged form 52 (one individual) to 0 (several individuals), with AM = 9.94, and SD = 11.57. The results of this second evaluation agreed significantly with the results of the ratings (r= .63, p <.001.)

## Use of Results

Individual results of the ratings process were mailed to the officers' homes. Despite the fact the one-third of the individuals were advised that their ratings placed them among the bottom third of rates, there seemed to be few complaints about the process or the results. The results were presented and discussed with the Chief. Since that time (almost two years later), ten lieutenants and captains from the overall top third have been promoted, four from the middle third have been promoted, and none promoted from the bottom third. In addition, individuals' ratings in particular dimensions have been considered in assigning and training individuals.

## Descriptive Statistics

A frequency count of the reasons for not having a performance appraisal system. Almost half (42.4%) of the barriers were due to a lack of resources and/or a lack of expertise. Sixteen percent of the municipalities reported union resistance as a serious barrier. The remainder of the barriers were managerial opposition (13.6%), non-support of mayor (9.6%), and employee resistance (8.8%). Thus, about half (47.8%) of the total barriers entailed resistance by users (employees, managers, and unions).

Gary E. Roberts is an assistant professor of public administration at Florida International University. His research focus is on human resources management with special emphasis on performance appraisal. Mr. Roberts was previously employed by the Pittsburgh Department of Public Safety as a senior research analyst specializing in human resource issues for law enforcement and fire. He is currently conducting research in performance measurement, age discrimination and public management. Mr. Roberts received his Ph.D. degree from the University of Pittsburgh.

The most important barriers to performance appraisal system development. Lack of expertise and resources continued to be the two most important barriers, but user resistance (raters, employees, unions) declined significantly. The most common reason offered was a lack of resources with 16 of the 40 responses followed by a lack of expertise with seven (a total of 57.5%). Lack of support from the mayor was reported to be the most important barrier in seven instances. Union resistance was selected by five respondents followed by managerial opposition (two) and employee resistance (one). Thus, it appears that technical and resource issues are the main reasons for the absence of a performance appraisal system, rather than employee or managerial opposition. It would have been useful to have identified those municipalities that had a performance appraisal system and subsequently abandoned it. The barriers to adoption may have been different based upon actual experience. This lack of significance of user resistance may reflect that on a philosophical basis, both management and employees support the concept of performance appraisal. Recent surveys of governmental employees indicated strong support for the merit concept, and the necessity of linking performance with personnel decision making.

The second part of the analysis attempts to determine if there were any systematic differences in the adoption of performance appraisal systems by geographic region, type of government, state of labor relations, or turnover and absenteeism rates. There may be systematic differences in the adoption of managerial innovations by region. Cities in the west and south have lower rates of unionization, hence the degree of management flexibility may be higher. City manager governments are more likely to adopt innovative personnel and management practices. If labor relations are poor, it is unlikely that an appraisal system will be effective because of the importance of trust. Higher and lower absenteeism rates may be symptomatic of underlying dissatisfaction or distrust. City manager governments were more likely to have a performance appraisal system. Eighty-six percent of the city manager cities had an appraisal system, compared to 64.9 percent of the non-city manager governments.

There were significant differences by geographic region. Eighty percent of the pacific, western and southern states utilized a performance appraisal system. In contrast, the eastern states had an operating performance appraisal system in only 17 of 49 municipalities (34.7 percent). The results on absenteeism were not significant, but there were significant differences by turnover rates. Cities with lower turnover rates were less likely to have an appraisal system. This could reflect the fact that unionized municipalities in the east provide a high degree of job security, hence lower turnover. Wage rates are lower in the south and in the west, as well as the degree of unionization. The result could be more management flexibility in exercising power which may contribute to employee dissatisfaction causing higher rates of turnover.

## Discriminant Analysis

To more fully test these effects, a discriminate analysis was performed. Discriminate analysis is most appropriate with categorical dependent variables. Discriminate analysis attempts to identify the variables that differentiate one group from another. The two category dependent variable discriminant analysis is very similar to traditional linear regression. There were a variety of continuous, ordinal and nominal variables that were entered. They were population, budget size, union percentage, absenteeism, turnover, type of government and geographic location. The results of the discriminate analysis. There were four significant variables that differentiated cities with and without performance appraisal systems. Cities with performance appraisal systems were more likely to have higher turnover rates, higher absenteeism rates, a favorable labor relations climate, and be a non-eastern city. The standardized canonical discriminant function coefficients indicate that geographic region and turnover rates are the two most influential variables. The overall explanatory power of the model was not satisfactory, however. The eigenvalue equals .27 indicating low between group variance. The canonical correlation is only .46, indicating a weak correlation between the variables and the discriminant functions. The classification analysis indicates that the percentage of "grouped" cases correctly classified is 74.22 percent, which was little better than chance.

Chapter V

Data Analysis

The above results indicate that the process of evaluating the promotability of mid-level and higher officers was effective. Solid levels of reliability and validity were found in the evaluations. Certainly, the large number of raters per ratee and the large variance between ratees helped produced a high degree of reliability. The validity of the evaluations was supported by the high correlation between the two evaluation processes, the lack of seniority or higher-rank bias, and the lack of self-interested ratings by any one rank of officers. [79] The variance of ratings and nomination was definitely sufficient for administrative decisions regarding and training of the rates. Participants' reactions were mostly neutral to positive, and subsequent promotions have generally reflected individuals' results, Besides being effective, this process was also quickly and inexpensively developed, especially compared to multiple choice examination and assessment centers. Yet higher-officer and peer evaluations of promotability are seldom used. What are likely sources of resistance and actual limitations to their use? [100]

Officers may feel that peer ratings especially may simply be a popularity contest, a formalized "good-old-boy" process. However, at least two researchers have partialled "friendship" out of peer ratings and found that while this factor may well be included in the ratings, the friendship factor does not significantly bias the validity of peer ratings. some of the same bases for friendship are also helpful qualities for higher-level positions (interpersonal skills, personal credibility). [115] A more serious limitation may be a low number of raters. The 2000 survey of law enforcement agencies' promotion methods reported that the number of raters making performance or promotability ratings varies between one and four. To provide the needed reliability and credibility of ratings, one to four or more raters are needed. Those few agencies which do conduct promotability ratings usually have only higher officers provide the ratings. Adding peers as raters would substantially help increase the reliability of the ratings and provide more complete, credible ratings of individuals.[128] While the ratings by peers and ratings by higher officers generally agreed in this evaluation, the overall peer averages for seven individuals differed from their overall higher-officer averages by more than 0.5 points. Clearly, using the both sources of raters provided more complete appraisals of these individuals, and helped offset possible bias by one source or the other. [101]

A more nebulous limitation bay be the professionalism and climate of work relations in an organization. If administrators or employees perceive much distrust, "back-biting," and office politics operating in their agency, this climate itself may prevent administrators form even attempting promotability ratings. [109] However, the risk of attempting these ratings can be fairly low. If agencies are willing to try promotability ratings, it would seem wise to first attempt their use long with other promotion methods, and with a relatively low weight. If the ratings are effective, this weight would then be increased for subsequent promotion cycles.[120]

Promotability ratings certainly seem underused by public safety agencies. To rely completely on more contrived methods of assessing promotability, while ignoring observations of officers' ongoing behavior in actual work settings, seems unwarranted. Reliable, valid ratings can be obtained relatively quickly and inexpensively, and the results of these ratings will probably be more acceptable to agency administrators and employees than those of other promotion methods.

## Demographic Characteristics

Of the 314 responses that were received, 242 (77.1 percent) employ at least one non- public safety performance appraisal system in public safety organization.(10) Most municipalities (59.2 percent) administer one appraisal system covering a mean of 91 percent of all their non- public safety employees. The mean time that the system had been in operation is 5.8 years and the vast majority of municipalities (82.1 percent) evaluate their employees once a year.

The mean municipal population is 129,549. Approximately two-thirds (63.8 percent) of the responding municipalities are council-manager governments, 29.6 percent are mayor-council governments and 6.3 percent are classified as other types (commission). The sample is representative in terms of geographic region and population. The respondents are primarily Personnel managers at public safety organization (62.1 percent), personnel officers (22 percent) or training directors (2.2 percent). The remainder are city managers (2.9 percent) and administrative assistants (6.7 percent). The respondents are, on average, an experienced group with a mean of 6.7 years of service in public safety personnel administration.

In terms of educational credentials, 47 percent have earned a masters degree while an additional 37 percent have obtained a bachelors degree. The most common majors are public administration (34 percent), business administration (19.8 percent), and personnel/labor relations (10.9 percent). The largest percentage of the respondents (43 percent) have completed 5 or more college courses in personnel administration, twenty-eight percent have taken 3 to 4 courses, and 29 percent have completed two or fewer courses. Almost all of the respondents (94 percent) have completed seminars or training

sessions on personnel issues. Eighty-seven percent are members
of at least one personnel-related professional association, and
all report being a member of at least one professional
association (personnel or other).[110] One area for concern is the
lack of personnel coursework for a substantial minority of the
sample. Nearly a third of the respondents have little formal
coursework in personnel issues. In all probability, at least one
of the courses is an introductory course with only a general
orientation to performance appraisal.


<u>Measurement</u>

The respondent's task was to rate the importance of 45
separate appraisal system attributes in building an effective
performance appraisal system in public safety organization. A
four-point Likert scale was used with response choices ranging
from "absolutely essential" to "not important." Based upon the
author's evaluation of the appraisal system literature, 42 of
the 45 total attributes are deemed to be very important in the
development and administration of an effective system. The three
remaining items are deemed to be desirable, but not essential. A
response is scored as agreeing with the literature if the
attribute is judged to be very important or absolutely
essential. If the attribute is judged to be less than very
important, the respondent's motivation to implement the

attribute may be reduced significantly.[122] This is especially important in the development and implementation stage, as attributes judged less important may be eliminated as part of the bargaining process in order to employ characteristics that are deemed more essential. All other things being equal, the probability of a system being judged effective should increase as more of these attributes are employed. [124]

Many of the items appear to be axioms and promote a desirability bias, but these attributes are extremely important to the success of any system, irrespective of their level of generality/desirability. For example, top management support is essential for the success of most management practices.(11) The survey is not designed to assess opinions on the controversial attributes/aspects of the appraisal process, but is a vehicle for gauging agreement with the research literature on the attributes deemed to be clearly essential for the operation of an effective appraisal system. [152]

Overall Agreement Levels

Of the 42 attributes that researchers have identified as very important, respondents indicate substantial agreement with 30. Substantial agreement is deemed to exist if 75 percent or more of the respondents select the "very important" category. There is agreement (50 percent or more) over 41 of the 42 attributes. Each individual item will be depicted in a series of bar charts grouped in a figure organized by the relevant aspect of the appraisal process being discussed. [191]

## Participation, Goal Setting and Feedback

Participation in the appraisal process encompasses four areas, developing performance standards, creating the rating form, appraisal interview participation, and employee self appraisal. Employee participation has been linked with higher levels of performance appraisal system in public safety organization satisfaction, fairness, acceptance and trust.(12) During the appraisal interview, employees should be encouraged to provide input, present their opinions and be able to rebut rater feedback that they disagree with.(13) A useful complement to this process is to require the completion of a performance self-appraisal before the actual interview to better prepare the employee and to focus attention on employee strengths and weaknesses.(14)

There are some interesting findings in regards to attitudes toward participation ??. Each type of participation can be implemented independently, hence there are likely to be differences on their perceived utility and importance, and the data support this hypothesis. By a 78 percent to 52 percent margin, respondents judge participation by raters in the development of the system to be more essential or important than participation by employees. Supervisory participation in developing performance standards is rated at least very important by 93 percent of the respondents compared to 73 percent for employee participation. Having employees rate their own performance is deemed to be at least very important by only 26.6 percent of the respondents. The lack of acceptance for self appraisal is consistent with its low rate of utilization as fewer than five percent of public safetys employ a self rating system.(15) Ninety-four percent of the respondents did recognize, however, that participation in the appraisal interview is very important or absolutely essential.

Ideally, participation should be operationalized in all stages of the process to maximize acceptance, but the limited data indicates that the benefits of participation in developing the appraisal system decay over time and that developmental participation is not associated with perceived user acceptance or effectiveness.(16) Thus, developmental participation has initial benefits, but the long-run effectiveness of the system is determined by how the appraisal system is administered. [123]

Goal setting is a critical component of an effective performance management program. Setting specific and moderately difficult goals will result in higher levels of performance and increased levels of motivation.(17) Employee participation in the development of performance goals increases goal difficulty and subsequent performance.(18) Respondents recognize the significance of goal setting, with 93 percent reporting that specific goals are important and 70 percent agreeing that goals should be tailored to individual employees. Participation in setting goals is deemed to be at least very important by 87 percent of the personnel professionals.[153] Respondents do not recognize, however, that goals should be moderately difficult to assure maximum motivation, as only 44 percent rate moderately difficult goals as very important. The literature states that goals that are too easy will not motivate sufficiently, while difficult goals will frustrate employees and result in withdrawal or diminution of effort. Goals should be challenging with a moderate probability for accomplishment.(19)

Performance feedback is both an input and an output of the performance appraisal process. Effective performance appraisal requires regular, ongoing two-way communication between rater and ratee.(20) Yearly performance appraisal evaluations are no substitute for the essential day-to-day interaction and coaching that is characteristic of effective supervision and leadership. Raters must be skilled at presenting feedback in a timely, specific, behavioral, and non-threatening fashion.(21) Regular performance counseling sessions in conjunction with the formal appraisal interview provide additional coaching and guidance.

The data on feedback is consistent with the research literature. Personnel professionals agree that regular performance counseling sessions are at least very important (91 percent) and that frequent informal performance feedback is at least very important (90 percent). The respondents concur that general performance feedback is not very important (only 31 percent important or essential). General feedback has little to no remedial value in correcting substandard performance.

## Training and Performance Standards

Performance standards are the building blocks of the appraisal process. In terms of the development of the system, the first step is the production of performance standards based upon a comprehensive job analysis.(22) The standards should be specific, measurable, clear, and communicated explicitly both orally and in writing.(23) Both raters and ratees should participate in developing performance standards and the rating form.(24) The dimensions of work performance that are evaluated should be the most important for effective job performance.(25)

As reflected, personnel professionals clearly recognize the importance of specific performance standards, (90 percent) based upon job analysis (83 percent), that are communicated clearly to employees (95 percent) and in which the rating instrument evaluates the most critical dimensions of job performance (92 percent). [133]

A very common and serious error in performance appraisal system in public safety organization implementation is the lack of formal rater training.(26) Assuming that raters possess the requisite skills to successfully administer the appraisal process without formal training is very risky. In training sessions, raters should receive instruction on how to document performance, preferably through a diary, and to be cognizant of the existence of various heuristic and attributional biases that distort information search and decision making.(27)

Research in performance appraisal demonstrates that performance appraisal system in public safety organizations require top-level management support to be successful.(28) Top-level support and commitment is demonstrated by holding managers accountable for how well they administer their performance appraisal responsibilities and by providing comprehensive performance appraisal training. [115]

The importance of these process variables are generally recognized by personnel professionals. Eighty-three percent or more of the respondents agree that rater training, evaluating supervisors on how well they administer the performance appraisal process, upper-management support, documentation of performance ratings, and the ability of the rater to observe

employee performance are very important or essential. A smaller
majority (59 percent) recognize the importance of considering
external factors beyond the employee's control that influence
performance. The lesser degree of agreement is congruent with
attribution theory research, which states that outside observers
tend to impute personal characteristics (lack of ability or
effort) as the cause for poor performance and to discount
external factors.(29) A failure to take external factors into
consideration when rating performance engenders perceptions of
unfairness among employees. [114]

## Employee and Rater Acceptance

Employee attitudes toward the appraisal system will
ultimately determine the success or failure of the system.(30)
Rater commitment is essential for an effective system, and an
important component of supervisory commitment is a rating form
that is easy to understand and use.(31) Employee acceptance is
facilitated when the appraisal system provides employee growth
and development and improves employer-employee relations.(32)
Another vital factor is rater-ratee agreement on the definition
of good performance and how performance appraisal information is
interpreted.(33) Other factors that influence acceptance are an
absence of race and sex bias, conformance to equal employment
opportunity law and confidentiality of appraisal information.

The limited research on the role of acceptance indicates that higher levels of acceptance are associated with perceptions that the appraisal system enhances motivation and productivity.(34)

Employee acceptance is recognized as very important or essential by 89 percent ??. The results for the other acceptance components indicate that the majority recognize the importance of rater and ratee agreement in interpreting performance information (89 percent), agreement in defining good performance (77 percent), the ability of the system to provide employee growth and challenge (84 percent), the ability of the system to improve employee-supervisor relations (88 percent), and the confidentiality of the appraisal system (89 percent).

Ninety-eight percent of the respondents agree that supervisor commitment is very important or essential. A factor related to rater acceptance is an appraisal form that is easy to use and understand, and 91 percent report that ease of use and understanding is essential or important. A clear demarcation between the rating form and the appraisal process itself is critical, however. The process of performance appraisal is anything but "easy" as it requires a complex web of knowledge, skills and abilities including effective performance observation techniques, precise documentation skills, a full knowledge of the job, the ability to provide clear performance feedback, effective goal setting practices, and the ability to cultivate meaningful employee participation. A balance must be struck between complexity in the process and a reduction in unnecessary paperwork and cumbersome appraisal forms.

In order for a system to be perceived as fair, it must be free of bias, including race and sex bias. Eighty percent state that avoiding adverse impact on protected groups is essential or important, and 89 percent agree that it is essential or important to conform to EEO law. The fact that the rate of agreement is not higher is somewhat surprising, but may be related to the developmental orientation of some systems where appraisal information is used for feedback purposes only.(35)

This lack of concern with validity by a minority of respondents is illustrated by the question on the importance of the appraisal system's scientific soundness (reliability and validity). Twenty percent report that scientific soundness is somewhat important and three percent that it is of little or no importance. These results may indicate that some appraisal systems are designed for feedback purposes only, or alternatively, it may reflect a lack of understanding of the concept. Another explanation is that respondents may believe a formal validation strategy is inherently unfruitful because of the underlying subjectivity of supervisory evaluations. An artificial validity may have utility in the laboratory, but quickly loses relevance in the complexity of the real world.

The three desirable, but not essential items, are the ability to standardize performance ratings to enable comparisons of employees with different supervisors (52 percent essential or important), the comparison of employees across different jobs (50 percent essential or important) and to computerize performance appraisal information (31 percent essential or important). The ability to standardize information across raters and jobs is very difficult and not practical for most systems, and computer adaptability is only cost-effective and practical in large organizations with many job titles and occupants. There are definite limits to the utility of standardization, and taken to the extreme, a rigid standardization can actually reduce effectiveness. A rigid standardization can make it difficult to adapt to changing job requirements and thereby reduce managerial appraisal flexibility and discretion.

## Performance Appraisal Information

The final set of questions relate to one of the most important products of the performance appraisal process, information for personnel decision making ??. Performance appraisal system in public safety organizations produce information for both developmental and administrative purposes. There is considerable discussion and controversy within the field on the appropriateness of using appraisals for administrative versus developmental purposes. This survey is designed to assess opinions on which individual uses are appropriate, and not to discuss the merits of developmental versus administrative objectives. However, what can be inferred from the results is an endorsement of both developmental and administrative purposes. Respondents state that the most essential and important types of information are information for performance improvement (91 percent), documenting demotion or discharge decisions (86 percent), identifying employee training needs (77 percent), making merit pay decisions (76 percent), and selecting employees for promotion and transfer (66 percent). Fewer respondents deem information for validating the appraisal system (64.2 percent) and identifying organization-wide problems (52 percent) as important or essential.

The results clearly indicate that there is substantial agreement with the performance appraisal literature. For a small subset of the items, there are considerable discrepancies and it would be useful to identify the variables that are associated with differences in attitudes toward their importance.

## Exploration of Differences - Aggregate Scores

This section will attempt to ascertain if respondent characteristics are associated with attitudes toward the conclusions of the appraisal literature.(36) One obvious hypothesis is that respondents who have earned a degree in a personnel related area are more likely to have been exposed to the relevant literature. Even if the respondent lacks specific personnel coursework, higher levels of education should be associated with greater agreement because many of the principles (participation, goal setting, and feedback) are components of general management education.

Another critical factor in shaping attitudes is the tone and nature of the respondent's experience with the performance appraisal system in public safety organization. Personal experience may be a more significant factor than education in explaining variance in agreement with the performance appraisal literature. If respondents believe that performance appraisal is a valid process, they are more likely to attribute performance appraisal system in public safety organization failure to flawed implementation. Conversely, if Personnel managers at public safety organization attribute performance appraisal ineffectiveness to inherent limitations of the technique (in a sense a defacto "theory" failure) the perceived importance of the recommended attributes will be reduced.(37) Implementation failures can be caused by a lack of managerial support, an absence of training, the inability or unwillingness to hold managers accountable for how well they administer the appraisal process, a lack of acceptance, and low levels of user trust or motivation, among others. Respondents may still retain belief in the efficacy of the attributes if the requisite resources and skills are subsequently made available.

If respondents lack confidence in the conceptual foundations of performance appraisal, these attitudes are likely to persist over time. For example, participants who believe that job performance measurement is inherently subjective and that performance appraisal introduces a false degree of objectivity will exhibit little support for the concept.(38) An important area for future research is to identify the variables that may contribute to opinions on the sources and causes of performance appraisal failure. One obvious factor is respondent experience with their own performance appraisal system in public safety organization. Over time, sustained performance appraisal system in public safety organization failure may change attitudes over the efficacy of the process and the utility of specific attributes such as participation.

If the attributes are not in place in the operating system, experience may not be as important as general education because there is no experiential basis for evaluating effectiveness. The study is handicapped in this regard because there are no items that assess perceived causes of performance appraisal system in public safety organization ineffectiveness. A series of two-group discriminant analyses were completed to identify the characteristics differentiating respondents on their levels of agreement with the performance appraisal literature.(39) The discriminant analysis model is not very effective in explaining variance in agreement with the performance appraisal literature or in classifying cases.(40)

Even though the model as a whole is not significant, there are a number of findings that suggest areas for additional research. Performance appraisal agreement is associated with non-council-manager governments, lower levels of education, and less job experience. The lower level of agreement for council-manager governments is surprising. One potential explanation may relate to the degree of experience with performance appraisal. Council-manager governments are more likely to employ a performance appraisal system in public safety organization.(41) Hence, if the respondent's experiences with performance appraisal are predominately negative, this may enhance beliefs that the "theory" proposed by the literature is flawed. This hypothesis is supported by the fact that those with more experience are less likely to agree with the performance appraisal literature. A more detailed analysis indicates that those with undergraduate degrees are more likely to agree with the literature than those with a masters. One possible explanation could be the interaction of education and experience. Those with masters degrees are more experienced, hence more opportunity to possess first-hand information about performance appraisal. On the other hand, it is plausible that the results reflect the rate of dissemination of performance appraisal research. Respondents who are less experienced are more likely to be recent graduates and to have been exposed to the latest performance appraisal literature.

The second part of the analysis was to examine differences of opinion over the utility of individual attributes. Substantial disagreement was reached when less than 75 percent of the respondents selected the attribute as being absolutely essential or important (except for the three that were considered only somewhat important). Each of these items were dichotomized and subjected to separate discriminant analyses employing the same model as was applied to the aggregate results. Twelve items received less than 75 percent agreement with 3 related to participation, 3 to goal setting, 1 to feedback, 3 addressed the types of information produced, 2 targeted validation issues, and 1 measured the importance of taking into consideration factors beyond the employee's control that influence performance.

Due to space limitations the individual discriminant analyses are not shown. Four of the 12 discriminant analyses are not significant overall. Even for those that attain significance, the amount of variance explained is very low, and the independent variables contribute little additional information in accurately classifying group membership. There are also no patterns or consistencies within the independent variables that offered insight in explaining disagreement with the literature. Thus, there appears to be no common factor that differentiates respondents on their level of agreement with the literature.

The analysis shows that Personnel managers at public safety organization exhibit a high level of agreement with the performance appraisal literature. This agreement encompasses the important attributes of participation, goal setting, feedback, information validity, employee and managerial acceptance, fairness, and organizational commitment. Thus, on at least a conceptual level, this agreement bodes well for the design and operation of performance appraisal system in public safety organizations. However, there is a major gulf between agreement with the literature and the actual design and administration of the appraisal system. Knowledge and agreement with the literature is only a necessary, but not a sufficient condition, for appraisal system effectiveness. There are a few areas for concern, however.

Personnel professionals do not value employee participation in developing the appraisal system as highly as rater participation. Participation in system development enhances employee understanding of job requirements, develops a consensus on what aspects of the job are important and how performance is to be measured. Participation in system development can increase employee commitment by transmitting a message that the organization values the employee's input and that the employee has a vested interest in the success of the system.(42) Self appraisal can be a very useful developmental tool, and has been shown to improve the appraisal process by enhancing the employee's understanding of the job, aids in the identification of performance problems, and provides for a more consensual and less defensive appraisal interview.(43) In terms of performance goals, it is often important to tailor goals to the individual job. "One size fits all" is inappropriate in the appraisal of job performance. Secondly, effective goal setting requires developing moderately difficult goals that are challenging but not impossible.

It is troubling that a small minority of respondents do not recognize the importance of avoiding adverse impact on protected groups. This lack of concern runs counter to established practice, as any performance appraisal system in public safety organization that is used for decision making must be assessed for adverse impact.(44) In explaining differences in the degree of agreement with the literature, the available data explains only a small percentage of the variance. Attitudes toward the literature are likely to be shaped by a complex web of variables related to education, training, personality, organizational climate, and experience with the performance appraisal system in public safety organization. The data suggest that those respondents who are more experienced with performance appraisal are less likely to agree with the literature. This may reflect dissatisfaction with the operating performance appraisal system in public safety organization. However, on an objective basis, there is widespread consensus on the critical attributes that are essential for the development and administration of an effective system.

Additional research should be undertaken to examine the attitudes of participants in terms of the perceived effectiveness of the appraisal process and the factors that enhance or detract from it. If participants are dissatisfied with the appraisal process, is this dissatisfaction attributable to perceived implementation failure or to the inherent limitations of the technique? Clearly performance appraisal system in public safety organizations are congruent with widely held values that emphasize merit principles and individual employee responsibility. Performance appraisal system in public safety organizations are compatible with the current societal and political culture that emphasizes the importance of making government (and all organizations) more efficient and effective.(45) The consistent support for performance appraisal, in spite of the considerable dissatisfaction with the process, indicates that there is an inherent faith in the concept and the values that underlie it.(46) This may explain the durability of performance appraisal and the continual attempts to improve the process. In all fairness, performance appraisal system in public safety organizations should change as technology, work processes, skills and abilities change. Performance appraisal is an ongoing process that requires elaboration and adaptation to changing circumstances.

If respondents attribute failure to a flawed conceptual foundation, the logical response will be to go through the motions, to create a performance appraisal "Potemkin Village" that satisfies legislative, administrative or press mandates, but that has little influence on employee behavior and performance. It is the authors experience that many appraisal systems are "paper tigers" that have little organizational relevance. What is required is additional research to gauge the depth and scope of these perceptions of all participants in the appraisal process as they will influence employee and rater behavior.(47)

Chapter VI

Conclusion

It is clear that performance appraisal systems are a "tool" of choice in the effort to make government more productive and to improve the productivity of individual employees. Performance appraisal is not universal, and it is clear that a municipality can operate without one. This research has attempted to identify the variables that distinguish which cities adopt a performance appraisal system from those that do not and the reasons why. An effective appraisal system can make a significant contribution to personnel management if the proper investment in time and resources is made. However, because resources are in short supply in these times of fiscal distress, it will be more difficult to develop and maintain a functional performance appraisal system.

In summary, it appears that personnel professionals are in agreement with the performance appraisal literature. This agreement extends to the basic processes of participation, goal setting, feedback, rater training, organizational commitment, rarer acceptance and ratee acceptance. Knowledge of effective practices is the first step in the process of developing a performance appraisal system in public safety organization that can meet individual and organizational goals. Additional research is necessary to further analyze the relationship between personnel manager attitudes/beliefs and the design and implementation of public safety performance appraisal system in public safety organizations.

An examination of the data indicates that lack of technical resources and an absence of expertise are the primary barriers to the adoption of a performance appraisal system. An effective performance appraisal system requires a considerable investment in time, energy and resources. Effective system development demands extensive participation by both raters and ratees. Time must be devoted to the development of the system including specific performance standards, clear performance goals, the rating form, and administrative procedures. There needs to be a systematic evaluation of the psychometric characteristics of the ratings, an identification of biased raters, and a monitoring of the quality of performance documentation. It is essential to have some type of grievance or appeal procedure to provide protection against abuse and bias. Employee and rater attitudes require monitoring to identify problem areas and strategies for improvement. Raters require initial training and follow-up training that should consist of clarification of the administrative procedures, instruction in diary-keeping, information processing strategies, feedback skills and goal setting techniques (Ilgen & Feldman; 2001, Bernardin & Beatty, 2001; Mohrman, Lawler, & Resnick-West, 1999). Too many municipal governments make a serious error by assuming that raters possess all of these abilities. Performance appraisal systems that are implemented without the proper technical and logistical support will in all probability, fail. The development and administration of a performance appraisal system requires specialized expertise and training that is not always available. Thus, it is better not to implement a performance appraisal system if the municipality lacks resources and expertise. A "paper" system destroys confidence in the operating system and damages the credibility and validity of performance appraisal in general.

Employee and rater attitudes play an important role. Other research has begun to document the importance of user attitudes in the ultimate effectiveness of the system. Employee and rater acceptance are critical surrogate measures for rating validity. A system that is designed according to the prescriptive and theoretical literature may be ineffective if the users perceive the system to be unfair, not useful, or too costly. It is important at the earliest stages to carefully assess the degree of trust in the appraisal process, as well as beliefs as to the availability of objective measures of performance, the ability and motivation of raters to rate accurately, and the overall climate of labor relations. Raters need to have confidence in their ability to rate accurately, be convinced of concrete benefits that outweigh the costs, and be reassured that the system will not disrupt work group harmony. Most traditional performance appraisal systems assume that individual employee performance is the proper unit of analysis. In reality, jobs vary on a continuum from complete independence to total integration within a team or group. Individual performance appraisal must be adopted to evaluate only those aspects of individual performance that are directly attributable to the employee, and design group appraisal methods for situations where the productive process is more integrated.

Unions have traditionally distrusted the ability and motivation of raters to fairly evaluate performance, and it is critical to gain their support in the developmental process. The importance of labor-management cooperation is underscored by the fiscal difficulties all levels of government are experiencing. Management must be careful not to use appraisal as a control device, while unions must re-orient their thinking and concentrate on means of increasing productivity and effectiveness.

The analysis clearly shows that municipalities in the west and south are more likely to have a performance appraisal system. This may be related to the higher degree of unionization in the northern states, but the variable for union representation was not significant. Not surprisingly, city manager governments are more likely to have a performance appraisal system. City manager governments typically create a "performance based culture" in contrast to the political culture of mayor-council governments.

Although the economic and societal benefit of efforts to clean up hazardous waste sites and consequently the utility of hazardous waste worker training are increasingly issues of public debate and commentary, an important message from this study is that progress is being made with respect to the training and performance of those responsible for cleaning up the nation's nuclear weapons complex. The extensive hazardous waste worker training at the Hanford nuclear waste site, the site of this overall study, occurs within a safety culture that values continuous safety education, the exhibition of safe work behaviors, and a concern for the health and well being of workers and the public. In our opinion, the Hanford safety culture undergirds the associated DOE contractor safety climates and the effectiveness of the present safety training and worker safety performance. The failure to make progress in these areas and the lack of a safety culture or safety climate have been singled out as primary precipitating factors in foreign nuclear waste disasters at Chernobyl and Tokaimura. In closing, this study provides a theoretical and empirical foundation for understanding general safety performance and the role of safety training histories that can be used in guiding research and practice efforts concerned with general safety performance.

Additional research is clearly warranted in identifying both the barriers to the adoption of a performance appraisal system and remedying the factors that contribute to ineffective performance appraisal systems. It would be ideal to identify clearly the total population of cities that lack a performance appraisal system, and to also identify cities that have eliminated performance appraisal systems and the reasons why.

## Recommendations

Listed below are recommendations for incorporating feedback from multiple sources into the performance appraisal process, along with a justification for each recommendation. Readers are encouraged to refer to the appendices for more detailed information on the basis for these recommendations.

During the first cycle, use the process only for managers. If the process works well and is accepted, expand the process to include all employees.

Unlike many employees, most managers had critical elements that were amenable to feedback from others. Planning, communication, and human resources management, all critical elements for management positions, can be evaluated by subordinates and/or peers.

This recommendation was supported by employee focus group participants and many managers who indicated that the process should be pilot tested on managers before rolling it out to all employees. In addition, all of the organizations surveyed were using the process for managers; only two had systems in place for nonmanagers. Thus, there was a wealth of information on using feedback from others in the performance appraisal process for managers that could guide the development and implementation of such a process in CEG.

Finally, the literature showed that older, more tenured, and higher educated employees, which managers typically are, generally are more resistant to this type of process. Working with them first should help surface any problems or concerns, making it a relatively smooth transition to expand the process to all employees.

Collect data from subordinates, peers, and self but not from customers.

The organizations surveyed and research literature examined fully supported collecting data from subordinates and peers. All indications were that subordinates and peers provide a unique perspective on job performance that the supervisor of that employee may not have. The research literature also showed that subordinate and peer ratings are reliable and valid predictors of job performance, and employees and managers are in favor of using them. Furthermore, other organizations have successfully developed and implemented feedback programs incorporating data from subordinates and peers with little resistance from those providing or receiving the data. While CEG managers and employees raised some concerns about using subordinate and peer data, these concerns could probably be alleviated by a well-developed implementation process and excellent up-front communication with those involved. For example, assurances that the anonymity of the raters will be maintained and that all data will be fed back to managers in an aggregated format may minimize any employee concerns that there could be repercussions for those providing their supervisors with poor ratings.

It was also recommended that individuals rate their own performance. While the literature pointed out several problems with self ratings, most organizations are in favor of using them. Furthermore, the purpose of incorporating other sources of information into the performance appraisal process is to provide individuals with a more complete picture of their performance. By comparing self ratings to subordinate and/or peer ratings, individuals should gain additional insight into their own strengths and weaknesses. In contrast, there was little information available that supported the use of customer feedback as a source of information in an individual's performance appraisal. Although a few papers have addressed the issue at a conceptual level, no research was uncovered that had examined the outcomes of incorporating customer feedback into the performance appraisal process.

All but one of the organizations surveyed recommended against using customer feedback in the process. The reasons given included:

* customers are better at evaluating products and services than evaluating individuals;

* any one employee typically has few customers, leading to concerns about the reliability and validity of the data and the anonymity of the source;

* customers often do not see the work of the employee, just the final products, and often do not know if or how the actions of the employee are limited by the rules, regulations, and resources of the organization;

* including customer feedback requires the development of a separate questionnaire, adding to the administrative burden of the process;

* customers asked to provide feedback for an employee may perceive that this employee is experiencing performance problems, which may impact the customer's perception of the employee in future interactions;

\* and, it is often uncomfortable for an employee or organization to ask a customer to provide feedback and for the customer to do so.

While we recommended against directly asking customers to evaluate an individual's job performance, we supported collecting customer feedback to help in the evaluation of CEG's products and services. To the extent that an individual participated in the development or implementation of the product or service, the customer information could be used in the evaluation of that individual. However, all sources examined recommended against directly polling customers about an individual employee.

Use the information for developmental, not evaluation, purposes.

Research and practice both overwhelmingly indicated that the data collection and feedback processes are most effective and efficient when performance ratings are collected for developmental rather than evaluation (e.g., to determine pay or promotions) purposes. When feedback data are used for evaluation purposes, employees often concentrate only on what needs to be done to achieve better ratings. When the data are used for developmental purposes, the employee can identify weaknesses and take action to improve upon those weaknesses. In addition, there are concerns about the accuracy of data collected for evaluation purposes. Two organizations surveyed conducted focus groups of employees who stated that because of fear of reprisal from their supervisors, they would give artificially high ratings to supervisors if the data were used to impact promotions and bonuses.

These findings are also supported by the literature which showed that employees and managers are much more favorable about receiving feedback from subordinates and peers when the ratings are used for developmental and not evaluation purposes. Furthermore, the research showed that ratings collected for evaluation purposes are more lenient, less reliable, less valid, and contain more halo (i.e., all dimensions are given similar ratings) than ratings made for developmental reasons. Data should be collected through a written questionnaire, with space made available for respondents to provide written comments.

The only two methods organizations surveyed used to collect subordinate and peer data were questionnaires and interviews. Questionnaires that allow respondents to provide ratings and written comments provide rich and useful data in a much more efficient and less costly manner than do interviews. In addition, the data collected on a questionnaire can be easily summarized and fed back to employees, ensuring the anonymity of respondents. Current practice suggested that the questionnaire be kept as short and as simple as possible. We suggested making the questionnaire approximately 30 items in length, with twenty of these items a set of core items used across CEG. The remaining items could be developed by individual offices in CEG, allowing for feedback in areas that are specific to the mission of the office. While the questions developed would be based, to some extent, on the performance appraisal elements for employees and managers, the elements themselves would not be changed. A consultant should be hired to analyze the data and develop feedback reports. If it is not feasible to hire a consultant, the process should be administered internally, with necessary training from PRD.

The research literature and organizations surveyed indicated that confidentiality is a critical issue that impacts the success or failure of such projects. Hiring a consultant ensures that nobody in CEG will have access to the feedback data of any other employee. About half of the organizations surveyed indicated that they use a consultant to analyze the feedback data and to generate individualized feedback reports.

The other half of the organizations surveyed stated that their human resources function, often their personnel department, maintains responsibility for analyzing the data and generating reports. This is often accomplished through the use of standardized computer programs that automatically analyze the data and develop the individualized feedback reports. These computer programs can be developed in-house or by a consultant.

The participating managers and their supervisors should select raters, and the managers should be responsible for distributing the questionnaires to those raters. The research literature and organizations surveyed indicated that if left to select respondents on their own, employees will often select only those who are likely to provide them with positive ratings. Individuals should work with their supervisors to identify respondents who are in the best position to rate their performance. This would include raters who have had frequent opportunities to observe the individual's performance. The supervisor could also help to ensure that the friendship factor is minimized.

In addition, the literature suggested that at least three raters should provide ratings for each individual. Including ratings from three or more raters improves the reliability and validity of the ratings, reduces the amount of bias in the ratings, and helps to maintain the anonymity of the raters. Finally, to minimize the cost of contracting with a consultant or the administrative burden on the personnel group, questionnaires should be provided to the manager, who is then responsible for distributing the questionnaires to the raters. Current practice indicated that such contact between the manager and the rater allows the manager to show that he or she supports the project and values the time and effort of the respondent. This, in turn, helps to ensure a high response rate.

In the first cycle, the feedback reports should compare self and other ratings for the manager to the average ratings for all managers at his or her organizational level. In future cycles, a comparison should also be made to the previous year's ratings for the manager. Organizations surveyed and past research placed the confidentiality of the ratings for an individual above all else. Because some offices had few managers at an organizational level, comparisons would have to be made at a broad enough level to ensure that an individual could not determine or reasonably guess someone else's ratings just by knowing his or her own ratings. Thus, it was recommended that the ratings of the manager should be compared to the average of all managers at the same organizational level.

Because the emphasis of the process was on self-improvement, it was also recommended that, in future years, the manager's current year ratings should be compared to his or her previous year's ratings. This would quickly allow the person to see areas of improvement from year to year. Feedback reports should be provided only to the manager for whom the data were collected and his or her immediate supervisor. Although some organizations provided feedback only to the employee on whom the ratings were made, the most successful programs also provided feedback to the individual's supervisor. The information was not used by the supervisor for evaluation; however, providing the feedback to someone else in the organization made the individual more responsible for taking action on areas in need of improvement.

Each manager and his or her supervisor should review the information together and identify strengths and developmental opportunities. They should form an action plan outlining specific activities the manager will undertake to improve upon weaknesses. Successful organizations indicated that the feedback data are most valuable when used in the development of action plans. According to the literature, action plans should be developed as soon after the feedback reports are received as possible. Acting upon the data quickly increases the likelihood that follow through will occur.

Finally, the organizations surveyed indicated that individuals typically focus their developmental activities in no more than three to five areas. Employees focusing on more than three to five areas tend to feel overwhelmed by the amount and complexity of data, and tend to not allocate sufficient time in any one area to make an impact on the quality or quantity of work in that area. CEG should provide the managers with the resources and support to improve on developmental areas.

The organizations surveyed indicated that it is worse to provide feedback from subordinates and peers and then not provide resources to improve than to not provide feedback at all. During or shortly after the meeting between the manager and his or her supervisor, the supervisor should ensure that support is available to allow the manager to work on developmental areas. During the course of the year, the supervisor should ensure that the manager has time to complete the developmental activities. The organizations also stated that in addition to formal training courses, other less costly alternatives can be used. For example, managers weak in certain areas can meet with or shadow managers strong in these areas. The feedback process, if successfully implemented during the first cycle, should be conducted annually. About half of the organizations surveyed conducted the feedback process just one time, usually to provide feedback to a small group of managers with communication or interpersonal difficulties. However, the other half of the organizations indicated that the process is most valuable when it is implemented on an annual or biennial basis. We recommended that the feedback process be conducted annually; however, developmental ratings should not be made at the same time as administrative ratings to avoid the appearance that they are being used for evaluation purposes. One suggestion was to use the midcycle review as an opportunity for providing developmental feedback to employees.

Upper management should demonstrate support for the process.

Each organization that indicated its feedback process was successful stated that upper management demonstrated support for the process. In fact, in two of the most successful organizations, the process began with a pilot test by executives. Once the executive teams felt comfortable with the feedback process, it was given to all managers in the companies. The process should be preceded by a memorandum from upper management to all employees explaining the purpose of the program, how it will be administered, and what will be done with the data. The memorandum should also ensure the confidentiality of the data and the anonymity of the respondents and stress that the process is being undertaken for developmental and not evaluation purposes. Finally, the memorandum should indicate the name and telephone number of a contact person who can be called should questions or concerns arise. If the feedback process is successful for managers, it could be expanded to all employees.

Summary

This paper presented recommendations for using 360 [degrees] feedback in performance appraisal. Although we did not recommend including customer evaluations or replacing individual with group appraisals, as called for by Deming and others in the TQM movement, we did incorporate procedures that allow for peer input and employee involvement. For example, we recommended including peer evaluations in the performance appraisal process. This allows coworkers to rate individuals on how much they cooperate and contribute to the group's performance. Being rated by one's coworkers should help to reinforce the importance of teamwork, which is one of the basic principles of TQM. Similarly, having employees rate their managers should make managers more aware of their own progress toward putting TQM principles, such as empowerment and team building, into practice.

We also recommended providing managers and employees with training opportunities for areas identified as needing improvement. Although all CEG employees are trained in TQM principles and practices, both managers and employees will need additional training to ensure that they have the skills and abilities necessary to operate in the rapidly changing work environment of the public safety organizations. Finally, we called for a strong commitment from management at all levels. Management in CEG has already taken many important steps to introduce TQM principles into its day-to-day operations. The recommendations presented in this paper illustrate one way that the performance appraisal process can be changed to promote

teamwork, employee involvement, and other TQM principles.

Endnotes

1. Anderson JR. (2002). Acquisition of cognitive skill. Psychological Review, 89, 369-406.

2. Anderson JR. (2003). Cognitive psychology and its implications (2nd ed.). New York: Freeman.

3. Austin JT, Villanova P. (2002). The criterion problem: 1917-2002. Journal of Applied Psychology, 77, 836-874.

4. Baldwin TT, Ford JK. (2000). Transfer of training: A review and directions for future research. PERSONNEL PSYCHOLOGY, 41, 63-105.

5. Borman WC, Hanson MA, Oppler SH, Pulakos ED, White LA. (2003). Role of early supervisory experience in supervisor performance. Journal of Applied Psychology, 78, 443-449.

6. Borman WC, Motowidlo SJ. (2003). Expanding the criterion domain to include elements of contextual performance. In Schmitt N, Borman WC (Eds.), Personnel selection in organizations (pp. 71-98). San Francisco: Jossey-Bass.

7. Borman WC, White LA, Pulakos ED, Oppler SH. (2001). Models of supervisory job performance ratings. Journal of Applied Psychology, 76, 863-872.

8. Campbell JP. (2000). Modeling the performance prediction problem in industrial and organizational psychology. In Dunnette MD, Hough LM (Eds.), Handbook of industrial and

organizational psychology: 2nd ed., (pp. 687-732). Palo Alto, CA: Consulting Psychologists Press.

9. Campbell IP, McHenry JJ, Wise LL. (2000). Modeling job performance in a population of jobs. PERSONNEL PSYCHOLOGY, 43, 313-333.

10.    Cantor JA. (2002). Evaluation of human performance in critical-skills occupations: Criteria and issues. Performance Improvement Quarterly, 5, 3-15.

11.    Carroll JS. (1998). Safety culture as an ongoing process: Culture surveys as opportunities for enquiry and change. Work & Stress, 12, 272-284.

12.    Cheyne A, Cox S, Oliver A, Tomas JM. (1998). Modeling safety climate in the prediction of levels of safety activity. Work & Stress, 12, 255-271.

13.    Chhokar JS. (2000). Behavioral safety management. Vikalpa, 15, 15-22.

14.    Cohen A. (1999). Factors in successful occupational safety programs. Journal of Safety Research, 9, 168-178.

15.    Farrell JN, McDaniel MA. (2001). The stability of validity coefficients over time: Ackerman's (2000) model and the general aptitude test battery. Journal of Applied Psychology, 86, 60-79.

16.     Flin R, Mearns K, O'Connor P, Bryden R. (2000).

Measuring safety climate: Identifying the common features.

Safety Science, 34, 177-192.

17.     Ford JK, Quinones MA, Sego D, Sorra J. (2002). Factors

affecting the opportunity to perform trained tasks on the

job. PERSONNEL PSYCHOLOGY, 45, 511-527.

18.     Griffin MA, Neal A. (2000). Perceptions of safety at

work: A framework for linking safety climate to safety

performance. Journal of Occupational Health Psychology, 5,

347-358.

19.     Hofmann DA, Jacobs R, Landy R. (1997). High

reliability process industries: Individual, micro, and

macro organizational influences on safety performance.

Journal of Safety Research, 26, 131-149.

20.     Hofmann DA, Stetzer A. (1998). A cross-level

investigation of factors influencing unsafe behaviors and

accidents. PERSONNEL PSYCHOLOGY, 49, 307-339.

21.     Hofmann DA, Morgeson FP, Gerras SJ. (2001). Climate as

a moderator of the relationship between LMX and content

specific citizenship behavior: Safety climate as an

exemplar. Unpublished manuscript.

22.     Hunt ST. (1998). Generic work behavior: An

investigation into the dimensions of entry-level, hourly

job performance. PERSONNEL PSYCHOLOGY, 49, 51-83.

23.     Hunter JE. (2003). A causal analysis of cognitive
        ability, job knowledge, job performance, and supervisory
        ratings. In Landy F, Zedeck S, Cleveland J (Eds.),
        Performance measurement and theory (pp.257-266). Hillsdale,
        NJ: Erlbaum.

24.     International Atomic Energy Agency. (1998). Highlights
        of conclusions and recommendations of "International
        Conference: One Decade After ChemobyL" Vienna, Austria:
        IAEA.

25.     International Atomic Energy Agency. (1999). Report on
        the preliminary fact finding mission following the accident
        at the nuclear fuel processing facility in Tokaimura,
        Japan. Vienna, Austria: IAEA.

26.     International Brotherhood of Teamsters. (2003).
        International Brotherhood of Teamsters hazardous waste
        worker training program: Student manual. Washington, DC:
        International Brotherhood of Teamsters.

27.     International Union of Operating Engineers. (undated).
        International Union of Operating Engineers national HAZMAT
        program: Department of Energy hazardous waste operations
        and emergency response training guide. Washington, DC:
        International Union of Operating Engineers.

28.     Jacobs R, Hofmann DA, Kriska SD. (2000). Performance
        and seniority. Human Performance, 3, 107-121.

29.    Joreskog KG, Sorbom D. (1998). LISREL & User's reference guide. Chicago: Scientific Software International, Inc.

30.    Kahneman D. (1998). Attention and effort. Englewood Cliffs, NJ: Prentice-Hall.

31.    Kanfer R, Ackerman PL. (1999). Motivation and cognitive abilities: An integrative/aptitude-treatment interaction approach to skill acquisition. Journal of Applied Psychology, 74, 657-690.

32.    Komaki J, Heinzmann AT, Lawson L (2000). Effects of training and feedback: Component analysis of a behavioral safety program. Journal of Applied Psychology, 65, 261-270.

33.    Lee T, Harrison K. (2000). Assessing safety culture in nuclear power stations. Safety Science, 34, 61-97.

34.    Lingard H, Rowlinson S. (1997). Behavior-based safety management in Hong Kong's construction industry. Journal of Safety Research, 28, 243-256.

35.    Maier NRF. (1999). Psychology in industry. Boston: Houghton-Mifflin.

36.    McCormick EJ, Jeanneret PR, Mecham RC. (2002). A study of job characteristics and job dimensions as based on the Position Analysis Questionnaire (PAQ). Journal of Applied Psychology, 56, 347-368.

37.     McDonald N, Corrigan S, Daly C, Cromie S. (2000). Safety management systems and safety culture in aircraft maintenance organizations. Safety Science, 34, 151-176.

38.     McHenry JJ, Hough LM, Toquam JL, Hanson MA, Ashworth S. (2000). Project A validity results: The relationship between predictor and criterion domains. PERSONNEL PSYCHOLOGY, 43, 335-366.

39.     Motowidlo SJ, Van Scotter JR. (2002). Evidence that task performance should be distinguished from contextual performance. Journal of Applied Psychology, 79, 475-480.

40.     Murphy K. (2000, March 12). Radioactive waste seeps toward Columbia River. Los Angeles Times [Online]. Available: www.latimes.com/news/science/environ/2000312/t000023730.html

41.     National Ironworkers and Employees Apprenticeship Training and Journeyman Upgrading Fund. (undated). Hazardous materials training for ironworker contractors and emergency response training guide. Washington, DC: International Association of Bridge, Structural and Ornamental Iron Workers.

42.     Neal A, Griffin MA, Hart PM. (2000). The impact of organizational climate on safety climate and individual behavior. Safety Science, 34, 99-109.

43.     Office of Technology Assessment. (2003). Hazards ahead: Managing cleanup worker health and safety at the nuclear weapons complex (DHHS Publication No. OTA-BP-O-85). Washington, DC: U.S. Government Printing Office.

44.     Oil, Chemical, and Atomic Workers International Union-The Labor Institute. (undated). Emergency response and prevention workbook: Awareness level and operations level. New York: The Labor Institute.

45.     Pate-Cornell ME. (2000). Organizational aspects of engineering system safety: The case of offshore platforms. Science, 250, 1210-1217.

46.     Quinones MA, Ford JK, Teachout MS. (1997). The relationship between work experience and job performance: A conceptual and meta-analytic review. PERSONNEL PSYCHOLOGY, 48, 887-910.

47.     Reason J. (1998) Achieving a safe culture: Theory and practice. Work & Stress, 12,293-306.

48.     Rudmo T (2000). Safety climate, attitudes, and risk perception in Norsk Hydro. Safety Science, 34, 47-59.

49.     Sackett PR, Gruys ML, Ellingson JE. (1998). Ability-personality interactions when predicting job performance. Journal of Applied Psychology, 83, 545-556.

50.     Schafer JL. (1997). Analysis of incomplete multivariate data. London: Chapman & Hall.

51.     Schmidt FL, Hunter JE, Outerbridge AN. (2000). Impact
        of job experience and ability on job knowledge, work sample
        performance and supervisory ratings of job performance.
        Journal of Applied Psychology, 71, 432-439.

52.     Stout RJ, Salas E, Kraiger K. (1997). The role of
        trainee knowledge structures in aviation team environments.
        The International Journal of Aviation Psychology, 7, 235-
        250.

53.     Tesluk PE, Jacobs RR. (1998). Toward an integrated
        model of work experience. PERSONNEL PSYCHOLOGY, 51, 321-
        355.

54.     The United Brotherhood of Carpenters Health and Safety
        Fund of North America. (2002). Hazardous waste participants
        manual. Cincinnati, OH: Midwest Consortium for Hazardous
        Waste Worker Training.

55.     Vineberg R, Taylor EN. (2002). Performance in four
        military jobs by men of different aptitude (AFQT) levels:
        The relationship of AFQT and job experience to job
        performance. Alexandria, VA: Human Resources Research
        Organization.

56.     Weiss HM. (2000). Learning theory and industrial and
        organizational psychology. In Dunnette MD, Hough LM (Eds.),
        Handbook of industrial and organizational psychology: 2nd

ed. (Vol. 1, pp.171-221). Palo Alto, CA: Consulting

Psychologists Press.

57.     Zohar D. (2000). Safety climate in industrial

organizations: Theoretical and applied implications.

Journal of Applied Psychology, 65, 96-102.

58.     Zohar D. (2000). A group-level model of safety

climate: Testing the effect of group climate on

microaccidents in manufacturing jobs Journal of Applied

Psychology 85, 587-596.

59.     Davidson, N B (May, 1999). Performance rated to weigh

promotions in Portland F. D. Fire Engineering, 24-30

60.     Eisenberg T. (January, 2000). An examination of

assessment center results and peer ratings. The Police

Chief, 46-47.

61.     Hollander, E. P. (2000). The friendship factor in peer

nominations. Personnel Psychology 9, 435-447.

62.     Kane, J.S. and Lawler, E. (2000). Methods of peer

assessment. Psychological Bulletin, 85, 555-586.

63.     Lewin, A. Y., and Zwany, A. (1998). Peer nomination: a

model, literature, critique, and a paradigm for research.

Personnel Psychology, 29, 423-447.

64.     Love, K. G. (2000). Accurate evaluation of police

officer performance through the judgment of fellow

officers: Fact of fiction? Journal of Police Science and
Administration, 9, 143-148.

65.     Schmitt, N., Gooding, R.Z., Noe, R.A. and Kirsch, M.
(2001). Meta-analysis of validity studies published between
1964 and 2002, and the investigation of study
characteristics. Personnel Psychology, 37, 407-422

66.     Scott, W. R. (January, 2001). Police promotional
procedure. The Police Chief, 54-56.

67.     Sherman, G. A.(2001). Promotional survey. Unpublished
manuscript, Washington State Patrol.

68.     Weiss, J.G. (2000). Statistical summary of response of
149 police departments to 2000 survey of examination
practices for promotion of the ranks of police sergeant,
lieutenant, and captain. Unpublished manuscript,
Metropolitan Police Department, Government of the District
of Columbia.

69.     Ammons, D. N. "Executive Satisfaction with Managerial
Performance Appraisal in City Government," Review of Public
Personnel Administration 7(2000): 33-48.

70.     Ammons, D. N., and S. E. Condrey. "Performance
Appraisal in Local Government: Warranty Conditions," Public
Productivity and Management Review 14,3(2001): 253-266.

71.     Ammons, D. N., and A. Rodriquez. "Performance
Appraisal Practices for Upper Management in City

Governments," Public Administration Review 46,5(2000): 460-467.

72.     Bassett, G. A. and H. H. Meyer "Performance Appraisal Based on Self-Review," Personnel Psychology 21(1998): 421-430.

73.     Bernardin, H. J. and R. W. Beatty. Performance Appraisal: Assessing Human Behavior at Work. Boston: Kent Publishing Company, 2001.

74.     Bowman, J. S. "At Last, an Alternative to Performance Appraisal: Total Quality Management," Public Administration Review 54(2002): 129-136.

75.     Burke, R. J., W. Weitzel, and T. Weir. "Characteristics of Effective Employee Performance Reviews and Development Interviews: Replication and Extension," Personnel Psychology 31(2000): 903-919.

76.     Carroll, S. J. and C. E. Schneier. Performance and Review Systems: The Identification, Measurement, Development of Performance in Organization. Dallas: Scott, Foresman and Company, 2002.

77.     Cardy, R. L and G. H. Dobbins. Performance Appraisal: Alternative Perspectives. Cincinnati: South-Western Publishing Co., 2002.

78.      Cedarbloom, D. "The Performance Appraisal Interview: A Review, Implications and Suggestions," Academy of Management Review 7(2002): 219-227.

79.      Daley, D. "Pay for Performance, Performance Appraisal, and Total Quality Management," Public Productivity & Management Review 16(2002a): 39-51.

80.      Daley, D. "Performance Appraisal in North Carolina Municipalities," Review of Public Personnel Administration, 12(2002b): 32-50.

81.      Daley, D. Performance Appraisal in the Public Sector. Westport, Connecticut: Quorum Books, 2002c.

82.      Deming, W. E. Out of the Crisis. Cambridge, Mass.: MIT Center for Advanced Engineering Study, 2000.

83.      Dillman, D. A. Mail and Telephone Surveys: The Total Design Method. New York: Wiley-Interchange, 2000.

84.      England, R. F., and W. M. Parle. "Nonmanagerial Performance Appraisal Practices in Large American Cities," Public Administration Review 7,6(2000): 498-504.

85.      Farah, J. F., J. D. Werbel, and A. G. Bedeian. "An Empirical Investigation of Self-Appraisal Based Performance Evaluation," Best Paper Proceedings of the Academy of Management (2000): 259-263.

86.      Feldman, J. M. "Instrumentation and Training for Performance Appraisal: A Perceptual-Cognitive Viewpoint,"

In K. M. Rowlan and G. R. Ferris, (Eds.), Research in Personnel and Human Resources Management, Greenwich, CT.: JAI Press, 2000.

87.    Folger, R. "Distributive and Procedural Justice in the Workplace," Social Justice Research 1(2000): 143-159.

88.    Fox, C. J. "Employee Performance Appraisal: The Keystone Made of Clay," In C. Ban & N.M. Riccucci (Eds.), Public Personnel Management: Current Concerns-Future Challenges. New York: Longman, 2001.

89.    Fox, S. F. "Professional Norms and Actual Practices in Local Personnel Administration," Review of Public Personnel Administration 13(2003): 5-28.

90.    Fox, S. F. and C. J. Fox. Merit Systems and Personnel Appraisal in Local Government (Baseline Data Report, vol. 22, no. 6) Washington, D.C.: International City Management Association, 2001.

91.    Gabris, G. T. "Can Merit Pay Systems Avoid Creating Discord Between Supervisors and Subordinates: Another Uneasy Look at Performance Appraisal," Review of Public Personnel Administration 7(2000): 70-89.

92.    Greenberg, J. "Organizational Performance Appraisal Procedures: What Makes Them Fair," In R.J. Lewicki et al. (Eds.), Research on Negotiation in Organizations: Vol. 1. (pp. 25-41). Greenwich, CT.: JAI Press, 2000a.

93.     Greenberg, J. (2000b). Determinants of perceived fairness of performance evaluations. Journal of Applied Psychology 71(2000b): 340-342.

94.     Greenberg, J. "The Distributive Justice of Organizational Performance Evaluations," In Justice in Social Relations, eds. H.W. Bierhoff et al., 337-351, New York: Plenum Press, 2000c.

95.     Greenberg, J. "Using Diaries to Promote Procedural Justice in Performance Appraisals," Social Justice Research 1(2000): 219-234.

96.     Greller, M. M. "Subordinate Participation and Reactions to the Appraisal Interview," Journal of Applied Psychology 60(1999): 544-549.

97.     Greller, M. M. "The Nature of Subordinate Participation in the Appraisal Interview," Academy of Management Journal 21(2000): 646-658.

98.     Henderson, R. "Developing Performance Measures Which are Consistent With the Mission of the Organization," In The Performance Appraisal Sourcebook Eds. L.S. Baird, R.W. Beatty and C.E. Schneier, 17-27, Amherst, Mass.: Human Resource Development Press, 2000.

99.     Harmon, M. M. and R. T. Meyer. Organization Theory for Public Administration. Glenview, IL.: Scott, Foresman and Company, 2000.

100.    Ilgen, D. R., and J. M. Feldman. "Performance Appraisal: A Process Focus," In Research in Organizational Behavior Vol. 5 Eds. L.L. Cummings and B.M. Staw, 141-197, Greenwich, CT.: JAI Press, 2003.

101.    Kane, J. S., and C. E. Lawler III. "Performance Appraisal Effectiveness. Its Assessment and Determinants," In Research in Organizational Behavior Vol. 1. Ed. B.M. Staw Greenwich, CT: JAI Press, 1998.

102.    Kanfer, R., Sawyer, J., P. C. Earley, and E. A. Lind. "Fairness and Participation in Evaluation Procedures: Effects on Task Attitudes and Performance," Social Justice Research 1(2000): 235-249.

103.    Kellough, J. E. and H. Lu. "The Paradox of Merit Pay in the Public Sector," Review of Public Personnel Administration 13(2003): 45-64.

104.    Lacho, K. G., G. K. Stearns, and M. Villere. "A Study of Employee Appraisal Systems of Major Cities in the United States," Public Personnel Management 8,2(1998): 111-125.

105.    Lacho, K. J., G. K. Stearns, and R. K. Whelan. "Performance Appraisal in Local Government: A Current Update," Public Productivity & Management Review 14,3(2001): 281-296.

106.    Landy, F. F., J. L. Barnes, and K. R. Murphy, K. R. "Correlates of Perceived Fairness and Accuracy of

Performance Evaluation," Journal of Applied Psychology 63(2000): 751-754.

107.    Latham, G. P., and T. W. Lee. "Goal Setting," In Generalizing from Laboratory to Field Settings. Ed. E. A. Locke, 101-117, Lexington, MASS: Lexington Books, 2000.

108.    Latham G. P. and K. N. Wexley. Increasing Productivity Through Performance Appraisal. Reading, Mass.: Addison-Wesley, 2000.

109.    Lissak, R. I. Procedural Fairness: How Employees Evaluate Procedures. Unpublished doctoral dissertation, University of Illinois, 2003.

110.    Locke, E. A. "Toward a Theory of Task Motivation and Incentives," Organizational Behavior and Human Performance 3(1998): 157-189.

111.    Lovrich, N., D. Bishop, R. Hopkins, and P. Shaffer. "Participative Performance Appraisal in a Municipal Setting: A Pre- and Post-Implementation Study," State and Local Government Review 15(2003): 24-31.

112.    Lovrich, N. P., R. H. Hopkins, P. L. Shaffer, and D. A. Yale. "Participative Performance Appraisal Effects Upon Job Satisfaction, Agency Climate, and Work Values: Results of a Quasi-Experimental Study in Six State Agencies" Review of Public Personnel Administration 1(2000): 51-73.

113.    Maroney, B. P., and R. Buckley. "Does Research in Performance Appraisal Influence the Practice of Performance Appraisal: Regretfully Not!" Public Personnel Management 21(2002): 185-196.

114.    Mohrman, A. M., S. M. Resnick-West and E. E. Lawler. Designing Performance Appraisal Systems. Jossey-Bass: San Francisco, 1999.

115.    Murphy, K. R., and J. N. Cleveland. Performance Appraisal: An Organizational Perspective. Boston: Allyn and Bacon, 2001.

116.    Pinder, C. C. Work Motivation. Glenview, Ill.: Scott, Foresman, 2001.

117.    Rainey, H. G. Understanding and Managing Public Organizations. San Francisco: Jossey-Bass, 2001.

118.    Roberts, G. E. "The Influence of Participation, Goal Setting, Feedback and Acceptance on Measures of Performance Appraisal System Effectiveness," Ph.D. diss., University of Pittsburgh, 2000.

119.    Roberts, G. E. "Linkages Between Performance Appraisal System Effectiveness and Rater and Ratee Acceptance: Evidence From a Survey of Municipal Personnel Administrators," Review of Public Personnel Administration 12,3(2002): 19-41.

120.    Roush, D. L., C. Curtis, H. A. Dershem, and N. P. Lovrich. "The Development of Behavior-Based Performance Appraisal Processes in Smaller Local Governments: Lessons From a Case Study," Public Productivity and Management Review 12(2001): 267-279.

121.    Schwabe, C. J. "Performance Appraisals in Local Government," Baseline Data Report Vol. 8, Washington, D.C.: International City Management Association, 2000.

122.    Silverman, S. B., and K. N. Wexley. "Reaction Employees to Performance Appraisal Interview as a Function of Their Participation in Rating Scale Development," Personnel Psychology 37(2001): 703-710.

123.    Steel, B. S. "Participative Performance Appraisal in Washington: An Assessment of Post-Implementation Receptivity," Public Personnel Management 14(2003):, 153-171.

124.    Steers, R. M. and T. W. Lee. "Facilitating Effective Performance Appraisals: The Role of Employee Commitment and Organizational Climate," In Performance Management and Theory, Eds. E. J. Landy and S. Zedeck, 75-85, 2003.

125.    Taylor, M. S., C. D. Fisher and D. R. Ilgen. "Individuals' Reactions to Performance Feedback in Organizations: A Control Theory Perspective," In Research in Personnel and Human Resources Management Volume 2. Eds.

K. Rowland and G. R. Ferris, 231-272, Greenwich, CT: JAI Press, 2001.

126.    Thayer, F. "Performance Appraisal and Merit Pay Systems: The Disasters Multiply," Review of Public Personnel Administration 7(2000): 36-53.

127.    Tubbs, M. E. "Goal Setting: A Meta-Analysis Examination of the Empirical Evidence," Journal of Applied Psychology 71(2000): 474-483.

128.    West, J. P. "City Personnel Management: Issues and Reforms," Public Personnel Management 13(2001): 317-334.

129.    Ammons, D. (2000). Executive satisfaction with managerial performance appraisal in city government. Review of Public Personnel Administration, 8, 33-48.

130.    Ammons, D., & Rodriguez, A. (2000). Performance appraisal practices for upper management in city governments. Public Administration Review, 46, 460-467.

131.    Bandura, A. (1999). Self-efficacy: Toward a unifying theory of behavioral change. Psychological Review, 84, 191-215.

132.    Bernardin, H., & Beatty, R. (2001). Performance appraisal: Assessing human behavior at work. Boston: Kent Publishing Company.

133.    Carroll, S., & Schneier, C. (2002). Performance and review systems: The identification, measurement,

development of performance in organizations. Dallas: Scott, Foresman and Company.

134.    Daley, D. (2001). Performance appraisal in North Carolina municipalities. Review of Public Personnel Administration, 11(3), 32-50.

135.    Dresang, D. (2000). Public personnel management. Reading, MA: Addison-Wesley.

136.    England, R. & Parle, W. (2000). Non-managerial appraisal practices in large American cities, Public Administration Review, 47, 498-504.

137.    Feldman, J. (2000). Beyond attribution theory: Cognitive processes in performance appraisal. Journal of Applied Psychology, 66, 127-148.

138.    Folger, R. (2000). Distributive and procedural justice in the workplace. Social Justice Research, 1, 143-159.

139.    Fox, C. (2001). Employee performance appraisal: The keystone made of clay. In Ban, C. & Riccucci, N. (Eds.), Public personnel management: Current concerns-future challenges (pp. 58-72) Longman: New York.

140.    Gabris, G. (2000). Can merit pay systems avoid creating discord between supervisors and subordinates: Another uneasy look at performance appraisal. Review of Public Personnel Administration, 7, 70-89.

141.    Greenberg, J. (2000a). Organizational performance appraisal procedures: What makes them fair. In R. J. Lewicki et al., (Eds.), Research on Negotiation in Organizations. (Vol-1, pp. 25-41). Greenwich, CT.: JAI Press.

142.    Greenberg, J. (2000b). Determinants of perceived fairness of performance evaluations. Journal of Applied Psychology, 71, 340-342.

143.    Greenberg, J. (2000c). The distributive justice of organizational performance evaluations. In H. W. Bierhoff et al., (Eds.), Justice in Social Relations, (pp. 337-351). New York: Plenum Press.

144.    Greenberg, J. (2000). Using diaries to promote procedural justice in performance appraisals. Social Justice Research, 1, 219-234.

145.    Ilgen, D., & Feldman, J. (2001). Performance appraisal: A process focus. In Research in Organizational Behavior, 5, 141-197.

146.    Lacho, K., Stearns, G., & Villere, M. (1998). A study of employee appraisal systems of major cities in the United States, Public Personnel Management, 8, 111-125.

147.    Latham, G., & Wexley, K. (2000). Increasing productivity through performance appraisal. Reading, Mass: Addison-Wesley.

148.    Laumeyer, J., & Beebe, T. (2000). Employees and their appraisal: how do workers feel about the company grading scale. Personnel Administrator, 33(12), 76-80.

149.    Lawler, E., Mohrman, A., & Resnick, S. (2000). Performance appraisal revisited. Organizational Dynamics, 20-35.

150.    Lovrich, N., Shaffer, R., Hopkins, D., & Yale, D. (2000). Public employees and performance Appraisal: Do their servants welcome or fear merit evaluation of their performance? Public Administration Review, 40(3), 214-222.

151.    Meyer, H. H. (1974). The pay-for performance dilemma. Organizational Dynamics, 3(1), 39-50.

152.    Mohrman, A., Resnick-West, S., & Lawler, E. (1999). Designing performance appraisal systems. San Francisco: Jossey-Bass.

153.    Nalbandian, J. (2000). Performance Appraisal: If only people were not involved, Public Administration Review, 41(3), 392-396.

154.    Norman, C. & Zawacki, A. (2001). Team appraisals-team approach. Personnel Journal, 70(9), 101-103.

155.    Norusis, M. J. (2003). SPSSX advanced statistics guide. New York: McGraw-Hill Book Company.

156.    Poister, T., & McGowan, R. (2001). The use of management tools in municipal government: A national survey. Public Administration Review, 44, 215-223.

157.    Rainey, H. (2001). Understanding and Managing Public Organizations. San Francisco: Jossey-Bass.

158.    Roberts, G. (2000). The influence of participation, goal setting and acceptance on measures of performance appraisal system effectiveness. Unpublished doctoral dissertation, University of Pittsburgh.

159.    Roberts, G. (2002) "Maximizing performance appraisal system acceptance: Perspective from municipal government personnel administrators," Paper submitted for publication.

160.    Schwabe, C. (2000). Performance appraisals in local government, Baseline Data Report, 8, Washington, D.C.: International City Management Association.

161.    Thayer, F. (2000). The presidents's management 'reforms': Theory X triumphant. Public Administration Review, 38(4), 309-314.

162.    Thayer, F. (2000). Performance appraisal and merit pay systems: The disasters multiple. Review of Public Personnel Administration, 7(2), 149-154.

163.    Bernardin, H.J. (2000). Subordinate appraisal: A valuable source of information about managers. Human Resource Management, 25, 421-439.

164.   Bernardin, H.J. & Beatty, R.W. (2001). Performance
       appraisal: Assessing human behavior at work. Boston: Kent.

165.   Cascio, W.F. (2000). Applied psychology in personnel
       management. Englewood Cliffs, NJ:Prentice-Hall, Inc.

166.   Cederblom, D. (2001). Promotability ratings: An
       underused promotion method for public safety organizations.
       Public Personnel Management, 20, 27-34.

167.   Cederblom, D. & Lounsbury, J.W. (2000). An
       investigation of user acceptance of peer evaluations.
       Personnel Psychology, 33, 567-579.

168.   Deming, W.E. (2002). Out of crisis. Cambridge: MIT.

169.   Die, A.H., Debbs, T., & Walker, J.L. (2000).
       Managerial evaluations by men and women managers. Journal
       of Social Psychology, 130, 763-769.

170.   Farh, J., Cannella, A.A., & Bedeian, A.G. (2001). Peer
       ratings: The impact of purpose on rating quality and user
       acceptance. Group and Organization Studies, 16, 367-386.

171.   Fox, S. & Bizman, A. (2000). Differential dimensions
       employed in rating subordinates, peers, and supervisors.
       Journal of Psychology, 122, 373-382.

172.   Freemantle, D. (2003). Incredible Customer Service.
       New York: McGraw-Hill.

173.     Harris, M.M. & Schaubroeck, J. (2000). A meta-analysis of self-supervisor, self-peer, and peer-supervisor ratings. Personnel Psychology, 41, 43-62.

174.     Kane, J.S. & Lawler, E.E., III (2000). Methods of peer assessment. Psychological Bulletin, 85, 555-586.

175.     Lewin, A.Y. & Zwany, A. (1998). Peer nominations: A model, literature critique, and a paradigm for research. Personnel Psychology, 29, 423-447.

176.     Love, K.G. (2000). Comparison of peer assessment methods: Reliability, validity, friendship bias, and user reaction. Journal of Applied Psychology, 66, 451-457.

177.     McEvoy, G.M. (2000). Using subordinate appraisals of managers to predict performance and promotions. Journal of Police Science and Administration, 15, 118-124.

178.     McEvoy, G.M. (2000). Public sector managers' reactions to appraisals by subordinates. Public Personnel Management, 19, 201-212.

179.     McEvoy, G.M. & Buller, P.F. (2000). User acceptance of peer appraisals in an industrial setting. Personnel Psychology, 40, 785-797.

180.     McEvoy, G.M., Buller, P.F., & Roghaar, S.R. (2000). A jury of one's peers. Personnel Administrator, 33, 94-101.

181.     Mohrman, A.M., Resnick-West, S.M., Lawler, E.E. III, Driver, M.J., Von-Glinow, M.A., & Prince, J.B. (1999).

Designing performance appraisal systems: Aligning appraisals and organizational realities. San Francisco, CA: Jossey-Bass Inc.

182.    Mount, M.K. (2001). Psychometric properties of subordinate ratings of managerial performance. Personnel Psychology, 37, 687-702.

183.    Mumford, M.D. (2003). Social comparison theory and the evaluation of peer evaluations: A review and some applied implications. Personnel Psychology, 36, 867-881.

184.    Norton, S.M. (2002). Peer assessments of performance and ability: An exploratory meta-analysis of statistical artifacts and contextual moderators. Journal of Business and Psychology, 6, 387-399.

185.    Riggio, R.E. & Cole, E.J. (2002). Agreement between subordinate ratings of supervisor performance and effects on self and subordinate satisfaction. Journal of Occupational and Organizational Psychology, 65, 151-158.

186.    Schmitt, N., Gooding, R.Z., Noe, R.A., & Kirsch, M. (2001). Meta-analysis of validity studies published between 1964 and 2002 and the investigation of study characteristics. Personnel Psychology, 37, 407-422.

187.    Tsui, A.S. & Ohlott, P. (2000). Multiple assessment of managerial effectiveness: Interrater agreement and

consensus in effectiveness models. Personnel Psychology, 41, 779-803.

188.   Vance, R.J., MacCallum, R.C., Coovert, M.D., & Hedge, J.W. (2000). Construct validity of multiple job performance measures using confirmatory factor analysis. Journal of Applied Psychology, 73, 74-80.

189.   Wexley, K.N. & Yukl, G.A. (2001). Organizational behavior and personnel psychology. Homewood, IL: Richard D. Irwin, Inc.

190.   Wexley, K.N. & Klimoski, R. (2001). Performance appraisal: An update. In Rowland, K.M. & Ferris, G.R. (Eds.), Research in personnel and human resources, (Vol. 2, pp. 35-79). Greenwich, CT: JAI Press.

191.   Wohlers, A.J. & Manuel, L. (1999). Ratings of managerial characteristics: Evaluation difficulty, co-worker agreement, and self-awareness. Personnel Psychology, 42, 235-261.

192.   Zedeck, S. & Cascio, W.F. (2002). Performance appraisal decisions as a function of rater training and purpose of the appraisal. Journal of Applied Psychology, 67, 752-758.

Notes

1. Jonathan West, "City Personnel Management: Issues and Reforms," Public Personnel Management, 13 (2001): 317-334.

2. S.F. Fox, "Professional Norms and Actual Practices in Local Personnel Administration," Review of Public Personnel Administration 13 (2003): 5-28.

3. W.E. Deming, "Out of the Crisis" in MIT Center for Advanced Engineering Study (Cambridge, Mass, 2000). F. Thayer, "Performance Appraisal and Merity Pay Systems: The Disasters Multiply," Review of Public Personnel Administration 7 (2000):36-53. C.J. Fox, "Employee Performance Appraisal: The Keystone Made of Clay," eds. In. C. Ban and N.M. Riccucci, Public Personnel Management: Current Concerns: Future Challenges (New York: Longman, 2001). D. Daley, "Pay for Performance, Performance Appraisal, and Total Quality Management," Review of Public Personnel Administration 16 (2002):39-51. J.S. Bowman, "At Last, an Alternative to Performance Appraisal: Total Quality Management," Public Administration Review 54 (2002): 129-136.

4. D. Daley, "Performance Appraisal in the Public Sector," (Westport, Connecticut: Quorum Books, 2002c).

5. S.J. Carroll and C.E. Schneier, Performance and Review Systems: The Identification, Measurement, Development of Performance in Organizations (Dallas: Scott, Foresman and Company, 2002).

6. R.J. Burke, "Characteristics of Effective Employee Performance Reviews and Development Interviews: Replication and Extension," Personnel Psychology 31 (2000): 903-919. J.S. Kane and C.E. Lawler, III," "Performance Appraisal Effectiveness. Its Assessment and Determinants," vol. 1 of Research in Organizational Behavior, ed. B.M. Straw (Greenwich, Ct.: JAI Press, 1998). R.M. Steers and T.W. Lee, "Facilitating Effective Performance Appraisals: The Role of Employee Commitment and Organizational Climate," in Performance Management and Theory, eds. E.J. Landy and S. Zedeck, (2003): 75-85. H.J. Bernardin and R. W. Beatty, Performance Appraisal: Assessing Human Behavior at Work (Boston: Kent Publishing Company, 2001). S.B. Silverman and K.N. Wexley, "Employees Reaction to Performance Appraisal Interviews as a Function of Their Participation in Rating Scale Development," Personnel Psychology 37 (2001): 703-710. A.M. Mohrman, S.M. Resnick-West and E.E. Lawler, Designing Performance Appraisal Systems (San Francisco: Jossey-Bass, 1999). G.E. Roberts, "The Influence of Participation, Goal Setting, Feedback and Acceptance on Measures of Performance Appraisal Effectiveness," Dissertation Abstracts International(Doctoral dissertation, University of Pittsburgh, 2000) 42:51-05a. K.R. Murphy and J.N. Cleveland, Performance Appraisal: An Organizational Perspective (Boston: Allyn and Bacon, 2001). D. Daley, Performance Appraisal in the Public

Sector(Westport, Ct.:Quorum Books, 2002c). R.L. Cardy and G.H. Dobbins, Performance Appraisal: Alternative Perspectives(Cincinnati: South-Western Publishing Co., 2002).

7. K.J. Lacho, G.K. Stearns and M. Villere, "A Study of Employee Appraisal Systems of Major Cities in the United States" Public Personnel Management8,2 (1998):111-125. C.J. Schwabe, "Performance Appraisal in Local Government," vol. 8 Baseline Data Report(Washington, D.C.: International City Management Association, 2000). D.N. Ammons and A. Rodriguez, "Performance Appraisal Practices for Upper Management in City Governments," Public Administration Review46,5(2000):460-467. D.N. Ammons, Executive Satisfaction with Managerial Performance Appraisal in City Government," Review of Public Personnel Administration 7(2000)33-48. R.F. England and W.M. Parle, "Nonmanagerial Performance Appraisal Practices in Large American Cities," Public Administration Review47,6 (2000):498-504. G.E. Roberts, "The Influence of Participation, Goal Setting, Feedback and Acceptance on Measures of Performance Appraisal System Effectiveness," Dissertation Abstracts International (Doctoral Dissertation, University of Pittsburgh, 2000).42:51-05a. S.F. Fox and C.J. Fox, "Merit Systems and Personnel Appraisal in Local Government," in International Management Association(Washington, DC, Baseline Data Report, vol.22, no.6, 2001). D. Daley, "Pay for Performance, Performance Appraisal,

and Total Quality Management," Public Productivity & Management Review 16 (2002a):39-51. D.N. Ammons and S.E. Condrey, "Performance Appraisal in Local Government: Warranty Conditions," Public Productivity and Management Review 14,3 (2001):253-266. D.L. Roush, C. Curtis, H.A. Dershem and N.P. Lovrich, "The Development of Behavior-Base Performance Appraisal Processes in Small Local Governments: Lessons from a Case Study," Public Productivity and Management Review 12 (2001):267-279. K.J. Lacho, G.K. Stearns and R.K. Whelan, "Performance Appraisal in Local Government: A Current Update," Public Productivity & Management Review14,3 (2001):281-296. B.P. Maroney and R. Buckley, "Does Research in Performance Appraisal Influence the Practice of Performance Appraisal: Regretfully Not!" Public Personnel Management21, (2002):185-196.

8. D.A. Dillman, Mail and Telephone Surveys: The Total Design Method(New York: Wiley-Interchange, 2000).

9. Schwabe, 2000. Fox & Fox, 2001.

10. Public safety appraisal systems were excluded from the analysis because public safety (police, fire, emergency medical services) personnel systems tend to operate with greater autonomy than other municipal departments, especially in larger cities.

11. H.G. Rainey, Understanding and Managing Public Organizations(San Francisco: Jossey-Bass, 2001).

12. N.P. Lovrich, R.H. Shaffer and D.A. Yale, "Participative Performance Appraisal Effects Upon Job Satisfaction, Agency Climate, and Work Values: Results of a Quasi-Experimental Study in Six State Agencies," Review of Public Personnel Administration 1 (2000):51-73. N. Lovrich, D. Bishop, R. Hopkins and P. Schaffer, "Participative Performance Appraisal in a Municipal Setting: A Pre and Post Implementation Study," State and Local Government Review15 (2003):24-31. R.I. Lissak, "Procedural Fairness: How Employees Evaluate Procedures" (Unpublished Doctoral Dissertation, University of Illinois, 2003). J. Greenberg, "Organizational Performance Appraisal Procedures: What Makes Them Fair," in Research on Negotiation in Organizations vol. 1, eds. R.J. Lewicki et al (Greenwich, CT.: JAI Press, 2000a):25-41. J. Greenberg, "Determinants of Perceived Fairness of Performance Evaluations," Journal of Applied Psychology 71 (2000b):340-342. J. Greenberg, "The Distributive Justice of Organizational Performance Evaluations," in Justice in Social Relations eds. H.W. Bierhoff et. al (New York: Plenum Press, 2000c):337-351. J. Greenberg, "Using Diaries to Promote Procedural Justice in Performance Appraisals," Social Justice Research 1 (2000):219-234. R. Folger, "Distributive and Procedural Justice in the Workplace," Social Justice Research 1 (2000):143-159. Lissak, 2003. Roberts, 2000. R. Kanfer, J. Sawyer, P.C. Earley and E.A. Lind, "Fairness and Participation

in Evaluation Procedures: Effects on Task Attitudes and Performance," Social Justice Research 1 (2000):235-249.

13. D. Cedarbloom, "The Performance Appraisal Interview: A Review, Implications and Suggestions," Academy of Management Review 7 (2002):219-227. M.M. Greller, "Subordinate Participation and Reactions to the Appraisal Interview," Journal of Applied Psychology, 60 (1999):544-549. M.M. Greller, "The Nature of Subordinate Participation in the Appraisal Interview," Academy of Management Journal (2000):646-658.

14. G.A. Bassett and H.H. Meyer, "Performance Appraisal Based on Self-Review," Personnel Psychology 21 (1998):421-430. J.F. Farah, J.D. Werbel and A.G. Bedeian, "An Empirical Investigation of Self-Appraisal Based Performance Evaluation," Best Paper Proceedings of the Academy of Management (2000):259-263. B.S. Steel, "Participative Performance Appraisal in Washington: An Assessment of Post-Implementation Receptivity," Public Personnel Management 14 (2003):153-171.

15. Schwabe, 2000. D. Daley, "Performance Appraisal in North Carolina Municipalities," Review of Public Personnel Administration 12, (2002b):32-50.

16. G.T. Gabris, "Can Merit Pay Systems Avoid Creating Discord Between Supervisors and Subordinates: Another Uneasy Look at Performance Appraisals," Review of Public Personnel Administration 7, 70 (2000):70-89. G.E. Roberts, "Linkages

Between Performance Appraisal System Effectiveness and Rater and Ratee Acceptance: Evidence from a Survey of Municipal Personnel Administrators," Review of Public Personnel Administration 12, 3(2002):19-41.

17. E.A. Locke, "Toward a Theory of Task Motivation and Incentives," Organizational Behavior and Human Performance 3 (1998):157-189. C.C. Pinder, Work Motivation (Glenview, IL.:Scott, Foresman, 2001). G.P. Latham and T.W. Lee, "Goal Setting," in Generalizing from Laboratory to Field Settings, ed. E.A. Locke (Lexington, MA: Lexington Books, 2000) 101-117. M.E. Tubbs, "Goal Setting: A Meta-Analysis Examination of the Empirical Evidence," Journal of Applied Psychology 71 (2000):474-483.

18. G.P. Latham and K.N. Wexley, Increasing Productivity Through Performance Appraisal, (Reading, Mass.: Addison-Wesley, 2000). Mohrman et al., 1999.

19. Locke, 1998.

20. Latham et al., 2000. H.J. Bernardin and R. W. Beatty, Performance Appraisal: Assessing Human Behavior at Work, (Boston, Kent Publishing Company, 2001). R.J. Burke, W. Weitzel and T. Weir, "Characteristics of Effective Employee Performance Reviews and Development Interviews: Replication and Extension," Personnel Psychology31 (2000):903-919. F.F. Landy, J.L. Barnes and K.R. Murphy, "Correlates of Perceived Fairness and Accuracy

of Performance Evaluation," Journal of Applied Psychology 63 (2000):751-754. M.S. Taylor, C.D. Fisher and D.R. Ilgen, "Individuals' Reaction to Performance Feedback in Organizations: A Control Theory Perspective," in Research in Personnel and Human Resources Management: vol 2, eds. K.Rowland and G.R. Ferris (Greenwich, CT: JAI Press, 2001)231-272.

21. Taylor et al., 2001.

22. Latham and Wexley, 2000.

23. Bernardin and Beatty, 2001.

24. Silverman and Wexley, 2001.

25. R. Henderson, "Developing Performance Measures Which are Consistent with the Mission of the Organization," in The Performance Appraisal Sourcebook eds., L.S. Baird, R.W. Beatty, & C.E. Schneier (Amherst, Mass.: Human Resource Development Press, 2000).

26. Maroney and Buckley, 2002.

27. Greenberg, 2000. D.R. Ilgen and J. M. Feldman, "Performance Appraisal: A Process Focus," in Research in Organizational Behavior, eds. L.L. Cummings and B.M. Staw, (Greenwich, CT.:JAI Press, 2003), vol. 5, 141-197. J.M. Feldman, "Instrumentation and Training for Performance Appraisal: A Perceptual-Cognitive Viewpoint," in Research in Personnel and Human Resource Management, eds. K.M. Rowlan and G.R. Ferris, (Greenwich, CT.:JAI Press, 2000)vol. 4, 45-99.

28. Mohrman et al., 1999.

29. Ilgen and Feldman, 2003.

30. S.J. Carroll and C.E. Scheier, Performance and Review Systems: The Identification, Measurement, Development of Performance in Organizations (Dallas: Scott, Foresman and Company, 2002). Bernardin et al., 2001.

31. Carroll and Schneier, 2002.

32. Carroll and Schneier, 2002. Roberts, 2000.

33. Roberts, 2000. Carroll and Schneier, 2002.

34. Roberts, 2002.

35. Bernardin and Beatty, 2001.

36. An overall agreement score was created by forming an additive scale for the 45 items. The items scoring scale ranged from 4 for "absolutely essential" to 1 for "not important at all." The maximum score attained was 172 and the minimum score was 62. An additive item Agreement Scale was created that exhibited an alpha of .89 with a mean score of 136.1 and a median of 137. A high level of agreement with the literature is represented by a score of 132 (an average of 3.0) or greater. Sixty-five percent of the sample had scores of 132 or higher. The 132 figure was obtained by requiring that the 42 items that were important should receive at least a very important rating (three points) for a total of 126. The three desirable, but not

essential items should receive a somewhat important rating (two points) for a total of 132.

37. M.M. Harmon and R.T. Meyer, Organization Theory for Public Administration, (Glenview, IL.: Scott, Foresman and Company, 2000).

38. C.J. Fox, "Employee Performance Appraisal: Keystone Made of Clay," in Public Personnel Management: Current Concerns Future Challenges eds. C. Ban and N.M. Riccucci (New York: Longman, 2001).

39. There were three broad types of variables that were included in the analysis. The first were respondent characteristics. They were the respondent's educational level, position (personnel versus other), the number of credits earned in personnel coursework, the respondent's degree area (personnel versus other), the number of credits earned in personnel coursework, the respondent's degree area (personnel versus other), and job tenure. Another group of variables related to organizational characteristics and included type of government, the level of unionization, and the state of labor relations. A third group of variables included community characteristics such as population (log converted) and geographic location. Respondents who scored less than 132 (less than 3.0 mean item score) on the Performance Agreement Scale were deemed to exhibit low agreement, while those above 133 were deemed to exhibit high agreement (above a

3.0 per item mean score). Approximately 35 percent of the respondents fell into the low agreement category.

40. The final discriminant model includes five variables. The Wilks' Lambda is .9467, indicating that there is very little difference in the group means for the independent variables. Thus, the majority of the variance is determined by non-measured variables. Correspondingly, the eigen value is .06 and the canonical correlation is .23 indicating that the between-groups sum of squares is low in comparison to the within-groups sum of squares. The classification model indicates that 63.6 percent of the grouped cases are correctly classified, which is equivalent to chance.

www.ingramcontent.com/pod-product-compliance
Lightning Source LLC
Chambersburg PA
CBHW081127170526
45165CB00008B/2582

9 781581 122695